TECHNICAL REPORT

The National Computer Security Survey (NCSS)

Final Methodology

Lois M. Davis, Daniela Golinelli, Robin Beckman,
Sarah K. Cotton, Robert H. Anderson, Anil Bamezai,
Christopher R. Corey, Megan Zander-Cotugno,
John L. Adams, Roald Euller, Paul Steinberg

Prepared for the Bureau of Justice Statistics

Safety and Justice

A RAND INFRASTRUCTURE, SAFETY, AND ENVIRONMENT PROGRAM

This report was prepared for the Bureau of Justice Statistics and conducted under the auspices of the Safety and Justice Program within RAND Infrastructure, Safety, and Environment (ISE).

Library of Congress Cataloging-in-Publication Data

National computer security survey : final methodology / Lois M. Davis ... [et al.].
    p. cm.
    Includes bibliographical references.
    ISBN 978-0-8330-4467-9 (pbk. : alk. paper)
    1. Computer crimes—United States. 2. Computer crimes—United States—Statistics. 3. Computer security—United States—Evaluation. I. Davis, Lois M.

HV6773.N38 2008
364.16'80973—dc22

2008018381

Published 2008 by the RAND Corporation
1776 Main Street, P.O. Box 2138, Santa Monica, CA 90407-2138
1200 South Hayes Street, Arlington, VA 22202-5050
4570 Fifth Avenue, Suite 600, Pittsburgh, PA 15213-2665
RAND URL: http://www.rand.org/
To order RAND documents or to obtain additional information, contact
Distribution Services: Telephone: (310) 451-7002;
Fax: (310) 451-6915; Email: order@rand.org

Preface

The Bureau of Justice Statistics (BJS) within the U.S. Department of Justice (DOJ), along with the U.S. Department of Homeland Security (DHS), conducted the National Computer Security Survey (NCSS), a nationally representative sample of businesses across 36 industry sectors. The RAND Corporation was selected to implement the survey. The survey collects data on the nature, extent, and consequences of computer-security incidents, with the goal of producing reliable national estimates of the incidence and prevalence of such incidents against businesses and their resulting losses. This technical report summarizes the methodology used in this study. This report was supported by BJS award 2004-BJ-CX-K024. The opinions, findings, and conclusions or recommendations expressed in this document are those of the authors and do not necessarily represent the official position or policies of DOJ.

This report will be of interest to researchers and analysts who are interested in understanding the methodology underlying the NCSS, including questionnaire design, survey fielding, sampling, and weighting methodologies for this large-scale survey. The report will also be informative for scholars who are interested in analyzing the NCSS public-use data file.

The RAND Safety and Justice Program

This research was conducted under the auspices of the Safety and Justice Program within RAND Infrastructure, Safety, and Environment (ISE). The mission of the ISE is to improve the development, operation, use, and protection of society's essential physical assets and natural resources and to enhance the related social assets of safety and security of individuals in transit and in their workplaces and communities. Safety and Justice Program research addresses occupational safety, transportation safety, food safety, and public safety—including violence, policing, corrections, substance abuse, and public integrity.

Questions or comments about this report should be sent to the project leader, Lois Davis (Lois_Davis@rand.org). Information about the Safety and Justice Program is available online (http://www.rand.org/ise/safety). Inquiries about research projects should be sent to the following address:

Greg Ridgeway, Acting Director
Safety and Justice Program, ISE
RAND Corporation
1776 Main Street
P.O. Box 2138

Santa Monica, CA 90407-2138
310-393-0411, x7737
Greg_Ridgeway@rand.org

Contents

Tables

Summary

The Bureau of Justice Statistics (BJS) in the Office of Justice Programs (OJP) of the U.S. Department of Justice (DOJ) conducted a pilot survey in 2001—the Computer Security Survey (CSS) pilot test—to collect information about computer infrastructure and security measures from businesses across a range of economic sectors, including finance, utilities, transportation, health care, telecommunications, petroleum mining and manufacturing, chemical and drug manufacturing, and Internet-service providers (ISPs).

Based on the pilot-survey results, BJS/DOJ, along with the U.S. Department of Homeland Security (DHS), decided to field a National Computer Security Survey (NCSS), a nationally representative sample of businesses across 36 industry sectors. The survey collected data on the nature, extent, and consequences of computer-security incidents. The goal was to produce reliable national estimates of the incidence and prevalence of computer-security incidents (including denial-of-service attacks and other incidents with serious consequences, such as fraud, embezzlement, theft of intellectual property [IP], electronic vandalism, or sabotage) against businesses, as well as businesses' resulting losses from such incidents. In 2004, RAND was selected to field the national survey. The survey itself was fielded in 2006; however, the data collected represent the experiences of companies in 2005.

This technical report details the methodology used to develop and field the NCSS as well as the sampling design and weighting methodology. The official survey results for the NCSS will be presented in a BJS report to be published in 2008. The NCSS is the first nationally representative survey on cyber security within the United States. More specifically, the survey was designed to collect data about the prevalence of different types of computer-security incidents and the monetary costs imposed by these incidents, across all industry sectors and especially for those that comprise the U.S. critical infrastructure. In addition, the NCSS provides information about what computer-security measures are currently in place, the reporting of computer-security incidents, and whom companies identify as being the perpetrators of such incidents.

Acknowledgments

The authors would like to thank the companies that have provided thoughtful information on their cyber-security postures and experiences with different types of computer-crime incidents. We also would like to thank the professional associations that lent their names in support of the NCSS and who helped get the word out to their industry sectors. Lastly, we wish to thank Ramona Rantala, Matthias Schonlau, and Patricia Gallagher, who reviewed and commented on drafts of this report.

Abbreviations

BJS	Bureau of Justice Statistics
CSS	Computer Security Survey
D&B	Dun and Bradstreet
DHS	U.S. Department of Homeland Security
DOJ	U.S. Department of Justice
DSP	data-safeguarding plan
D-U-N-S	data universal numbering system
ESS	effective sample size
FAQs	frequently asked questions
HSPC	Human Subjects Protection Committee
IP	intellectual property
ISAC	Information Sharing and Analysis Center
ISP	Internet-service provider
IT	information technology
MAR	missing at random
MCAR	missing completely at random
MNAR	missing not at random
MSI	Market Strategies International
NAICS	North American Industry Classification System
NCSD	National Cyber Security Division
NCSS	National Computer Security Survey
OJP	Office of Justice Programs
OMB	Office of Management and Budget

OPSCAN   optical scanning

PDF   portable document format

PGP   pretty good privacy

PIN   personal identification number

RFP   request for proposals

RMS   record-management system

SAQ   self-administered questionnaire

SRG   Survey Research Group

SUDAAN   survey-data analysis

CHAPTER ONE

Introduction

The Bureau of Justice Statistics (BJS) in the Office of Justice Programs (OJP) of the U.S. Department of Justice (DOJ) conducted a pilot survey in 2001—the Computer Security Survey (CSS) pilot test—to collect information about computer infrastructure and security measures from businesses across a range of economic sectors, including finance, utilities, transportation, health care, telecommunications, petroleum mining and manufacturing, chemical and drug manufacturing, and Internet-service providers (ISPs).

Based on the pilot-survey results, BJS/DOJ, along with the U.S. Department of Homeland Security (DHS), decided to field a National Computer Security Survey (NCSS), a nationally representative sample of businesses across 36 industry sectors. The survey collected data on the nature, extent, and consequences of computer-security incidents. The goal was to produce reliable national estimates of the frequency and prevalence of computer-security incidents (including denial-of-service attacks and other incidents with serious consequences, such as fraud, embezzlement, theft of intellectual property [IP], electronic vandalism, or sabotage) against businesses, as well as businesses' resulting losses from such incidents. In 2004, RAND was selected to field the national survey. The survey itself was fielded in 2006; however, the data collected represent the experiences of companies in 2005.

Objective

This technical report details the methodology used to develop and field the NCSS as well as the sampling design and weighting methodology. The official survey results for the NCSS will be presented in a BJS report to be published in 2008. The NCSS is the first nationally representative survey on cyber security within the United States. More specifically, the survey was designed to collect data about the prevalence of different types of computer-security incidents and the monetary costs imposed by these incidents, across all industry sectors and especially for those that comprise the U.S. critical infrastructure. In addition, the NCSS provides information about what computer-security measures are currently in place, the reporting of computer-security incidents, and whom companies identify as being the perpetrators of such incidents.

Organization of This Report

In this report, we have organized the description of the methodology used for the NCSS into five chapters. Chapter Two addresses frame definition, exclusions from the frame, and cleaning of the frame. Chapter Three presents the sampling methodology. Chapter Four presents the survey methodology. Chapter Five presents weighting methodology and nonresponse analysis, as well as examples of weighted estimates and standard errors using SUDAAN software.[1] Appendix A provides details of the range and edit checks. Appendix B provides a copy of the long form of the questionnaire.

---

[1]   SUDAAN stands for survey-data analysis.

CHAPTER TWO

Frame Definition

Rationale for Using Dun and Bradstreet

While the 2001 CSS pilot study used the U.S. Census Bureau's Business Register to derive its sampling frame, that database is not publicly available and so was unavailable to use for the NCSS. As an alternative, we selected Dun and Bradstreet (D&B®) as the database vendor of choice, given the complexity of content it can provide. Specifically, the D&B database has information on approximately 18.5 million active U.S. businesses on record. In comparison, InfoUSA's and Experian's databases contain only 13 million and 15 million businesses, respectively; in addition, we found that those databases had less depth of information on company hierarchy and corporate structure, or "family-tree" mapping, than D&B's database had. Also, D&B's database is continually updated, both manually and with automated data-collection tools. These updates are made from numerous sources, including public filings, global business-to-business transactions, direct investigations, Internet presence, and media reports. The National Science Foundation's *Information Technology Innovation Survey: Fall 2001*, which had similar sampling priorities as the NCSS survey, also used D&B's database (NSF, 2004).

D&B creates a nine-digit D-U-N-S® (data universal numbering system) for each company in its system as a unique way of tracking and identifying businesses through changes and updates. This number is retained for the life of a business, with no two businesses ever receiving the same D-U-N-S number. D-U-N-S numbers are not recycled; they follow a company from birth to growth and even death. Important to this study, the D&B database is able to reflect corporate hierarchy and corporate families based on each business' D-U-N-S number.

Dealing with Large Corporations and Their Subsidiaries

D&B's Strategic Marketing Records database served as the sampling frame for the NCSS. We received the data in February 2005 and drew the sample in November 2005, requesting at that time updated contact information.

The frame consisted of all companies operating in the United States with 10 or more employees. Given the large number of companies in the 2- to 9-employee size grouping and the small sampling fraction intended for this size group, we did not purchase the full set of records for the small companies. To include them in the sampling frame, we obtained counts by North American Industry Classification System (NAICS) codes, which were then used to group them into industries. The sampling allocation was done by industry, and the number of small companies allocated was converted back to counts per NAICS code. We then requested that D&B generate a random sample of small companies within each NAICS code based on these counts. The translation back to NAICS from industry was calculated by the percentage that a given NAICS code contributed to the total industry group. We purchased this stratified random sample of small companies. Companies in the D&B database with an employee count of one are largely nonemployers with just a single person operating the business and, thus, were not included in the sampling frame.

The sampling unit was a company, which we define as one or more business establishments under common ownership. Branches and divisions were not considered a company; however, subsidiaries of larger corporate trees were.

The sampling frame consisted of three types of companies:

1. companies comprised of one establishment (e.g., single-location companies)
2. companies that may have several establishments nationwide but only one company headquarters and that is not, in turn, owned by another company
3. companies that are part of a large conglomerate (i.e., subsidiaries of other companies, all of which form a pyramid with a final entity at the apex—what D&B refers to as the "ultimate" company).

From a sampling point of view, we define a company in the following way: Each single-location company or corporate headquarters of a multiestablishment company is treated as one sampling unit. Headquarters divide into two groups: (1) parent or ultimate headquarters and (2) subsidiary headquarters. We describe next how we dealt with companies that were part of large conglomerates.

The complexity of some corporate trees (category 3 in the list) was highlighted in the cognitive interviews we conducted. Information-security practices were found to vary considerably across companies, even though the companies might be part of the same corporate tree. The ultimate entity at the apex of a complex corporate tree was not always able to report about cyber security–related information for all its subsidiaries.

Further, analysis of the complex corporate trees included in D&B's database and informal discussions with company contacts indicated that there is wide heterogeneity in how IT is structured among these complex companies, which, in many instances, may limit the ultimate companies' information about their subsidiaries' experience with computer security incidents and the ultimates' ability to report for all their subsidiaries. In addition, companies that were part of a complex corporate tree might also belong to different industrial sectors (our data suggest that approximately 20 percent of corporate trees span two or more of the 36 industry sectors). Yet a breakdown by subsidiary was required to allow for industry-level estimates (i.e., each company, including subsidiaries, needed to be in its proper industry sector).

For complex corporate trees, we struck a compromise. Because IT systems and information-security practices may not be fully integrated within complex companies and because subsidiaries of complex companies may belong to different industries, it was decided to treat subsidiaries belonging to the same parent companies independently. In other words, if a company owned, for example, five other companies, the links among these six companies were erased and all six companies had a positive sampling probability. That is, each company part of a complex tree was treated as an independent sampling and reporting unit. This decision also was reached due to sample size and budget constraints. Given a fixed budget, another option would have been to reduce the sample size and use available resources to do more-intensive follow-up with the complex companies, allowing an entire corporate tree to be treated as a sampling unit.

In addition, the survey instructed the target respondent to report only for his or her domestic operations, including all divisions and locations and excluding subsidiaries. The NCSS also contained a check at the end to confirm that subsidiary data had been excluded. This was done because we recognized that some companies may be unable to report information independent of their subsidiaries. So two questions were added at the end of the survey to verify for whom the respondent was reporting and the number of subsidiaries for which they were reporting. If the respondent indicated that information for its subsidiaries was included, the respondent was asked to list which additional companies were included in their report.

Thus, in some cases, companies filled out the survey including information for subsidiary firms within their corporate structure. We determined this from the response to question 21: "Information includes this company and some subsidiaries." We obtained a list of these subsidiaries and compared these to the survey frame to determine whether any of the subsidiaries were part of the NCSS sample. If so, we included records for those subsidiaries in the database. We did not, however, fill in any of the fields for these subsidiary records, leaving these decisions up to the discretion of future analysts who may wish to analyze these data. Therefore, on the analytic file, all the fields on the subsidiary records are blank, except for the sampling strata and weight variables.

In using these data, analysts will need to make their own decisions whether, and how, to distribute the information on the parent-company record to the subsidiary records. Consider the following example: A parent company has three blank subsidiary records representing the companies in the sample frame for which the parent company provided data. If one assumes that the information was evenly distributed across the four companies (the parent company plus the three subsidiaries), then the counts of incidents detected, incidents reported, the dollar costs of the incidents, and so on, could be apportioned such that each company received 25 percent of the total. This is just one simple example of how the information supplied on behalf of subsidiary companies might be allocated to those companies.

Exclusions from the Frame

The NCSS focused on private-sector companies (whether domestically or internationally owned) that are operating in the United States. Companies were grouped into industries based on NAICS codes. A small number of codes were deemed ineligible and excluded from the sampling frame. In consultation with BJS, we excluded from the frame those companies that belonged to the public sector, the education sector, and private households.

A complete listing of the excluded codes and their definitions can be found in Table 2.1.

Table 2.1
Exclusions from the Sampling Frame: Ineligible NAICS Codes

NAICS Code	Description
491110	Postal Service
611110	Elementary and Secondary Schools
611210	Junior Colleges
611310	Colleges, Universities and Professional Schools
814110	Private Households
921110	Executive Offices
921120	Legislative Bodies
921130	Public Finance Activities
921140	Executive and Legislative Offices, Combined
921150	American Indian and Alaska Native Tribal Governments
921190	Other General Government Support
922110	Courts
922120	Police Protection
922130	Legal Counsel and Prosecution
922140	Correctional Institutions
922150	Parole Offices and Probation Offices
922160	Fire Protection
922190	All Other Justice, Public Order, and Safety Activities
923110	Administration of Education Programs
923120	Administration of Public Health Programs
923130	Administration of Human Resource Programs (except Education, Public Health, and Veteran's Affairs Programs)
923140	Administration of Veteran's Affairs
924110	Administration of Air and Water Resource and Solid Waste Management Programs
924120	Administration of Conservation Programs
925110	Administration of Housing Programs
925120	Administration of Urban Planning and Community and Rural Development
926110	Administration of General Economic Programs
926120	Regulation and Administration of Transportation Programs
926130	Regulation and Administration of Communications, Electric, Gas, and Other Utilities
926140	Regulation of Agricultural Marketing and Commodities
926150	Regulation, Licensing, and Inspection of Miscellaneous Commercial Sectors
927110	Space Research and Technology

Table 2.1—Continued

NAICS Code	Description
928110	National Security
928120	International Affairs

Industry Definitions

In agreement with the sponsor, we decided to define the industry sectors according to the 2002 NAICS groupings. Table 2.2 provides the specific NAICS codes assigned to each industry sector. Even though a company might have several NAICS codes depending on the lines of business in which it operates, it was decided to assign every company to the industry corresponding to its primary NAICS code.

Table 2.2
NAICS Codes That Defined Each Industry Sector

Industry Code	Industry	NAICS Code (2–6 digits)	NAICS Definition
1	Accommodations	721	Accommodation
2	Accounting & Bookkeeping	5412	Accounting, Tax Preparation, Bookkeeping, and Payroll Services
3	Administrative Support	56	Administration Support and Waste Management and Remediation Services
4	Advertising & Other Technical Services	5418	Advertising and Related Services
		5419	Other Professional, Scientific, and Technical Services
5	Agriculture	111	Crop Production
		112	Animal Production
6	Architecture, Engineering & Specialized Design Services	5413	Architectural, Engineering, and Related Services
		5414	Specialized Design Services
7	Arts & Entertainment	71	Arts, Entertainment, and Recreation
8	Business & Technical Schools	6114	Business Schools and Computer and Management Training
		6115	Technical and Trade Schools
		6116	Other Schools and Instruction
		6117	Education Support Services
9	Chemical & Drug Manufacturing	325	Chemical Manufacturing
11	Computer Systems Design	5415	Computer Systems Design and Related Services
12	Construction	23	Construction

Table 2.2—Continued

Industry Code	Industry	NAICS Code (2–6 digits)	NAICS Definition
13	Finance	521	Monetary Authorities – Central Bank
		522	Credit Intermediation and Related Activities
		523	Securities, Commodity Contracts, and Other Financial Investments and Related Activities
14	Food Services	722	Food Services and Drinking Places
15	Forestry, Fishing & Hunting	113	Forestry and Logging
		114	Fishing, Hunting, and Trapping
		115	Support Activities for Agriculture and Forestry
16	Health Care	621	Ambulatory Health Care Services
		622	Hospitals
		623	Nursing and Residential Care Facilities
17	Insurance	524	Insurance Carriers and Related Activities
		525	Funds, Trusts, and Other Financial Vehicles
18	ISPs [Internet-service providers] & Web Search Portals	518	Internet Service Providers, Web Search Portals, and Data Processing Services
19	Legal Services	5411	Legal Services
20	Manufacturing Durable	321	Wood Product Manufacturing
		327	Non-Metallic Mineral Product Manufacturing
		331	Primary Metal Manufacturing
		332	Fabricated Metal Product Manufacturing
		333	Machinery Manufacturing
		334	Computer and Electronic Product Manufacturing
		335	Electrical Equipment, Appliance, and Component Manufacturing
		336	Transportation Equipment Manufacturing
		337	Furniture and Related Product Manufacturing
		339	Miscellaneous Manufacturing

Table 2.2—Continued

Industry Code	Industry	NAICS Code (2–6 digits)	NAICS Definition
21	Manufacturing Non-Durable	311	Food Manufacturing
		312	Beverage and Tobacco Product Manufacturing
		313	Textile Mills
		314	Textile Product Mills
		315	Apparel Manufacturing
		316	Leather and Allied Product Manufacturing
		322	Paper Manufacturing
		323	Printing and Related Support Activities
		324121	Asphalt Paving Mixture and Block Manufacturing
		324122	Asphalt Shingle and Coating Materials Manufacturing
		324191	Petroleum Lubricating Oil and Grease Manufacturing
		324199	All Other Petroleum and Coal Products Manufacturing
		326	Plastics and Rubber Products Manufacturing
22	Mining	212	Mining (except Oil and Gas)
		213	Support Activities for Mining
23	Other Services	811	Repair and Maintenance
		812	Personal and Laundry Services
		813	Religious, Grant making, Civic, Professional, and Similar Organizations
24	Petroleum Mining & Manufacturing	2111	Oil and Gas Extraction
		32411	Petroleum Refineries
25	Picture & Sound Recording	5121	Motion Picture and Video Industries
		5122	Sound Recording Industries
26	Publications & Broadcasting	5111	Newspaper, Periodical, Book, and Directory Publishers
		5112	Software Publishers
		5151	Radio and Television Broadcasting
		5152	Cable and Other Subscription Programming
		5161	Internet Publishing and Broadcasting
		5191	Other Information Services
27	Real Estate	531	Real Estate

Table 2.2—Continued

Industry Code	Industry	NAICS Code (2–6 digits)	NAICS Definition
28	Rental Services	532	Rental and Leasing Services
		533	Lessors of Non-Financial Intangible Assets (except copyrighted works)
29	Retail	44	Retail Trade
		45	Retail Trade
30	Scientific Research & Development	5416	Management, Scientific, and Technical Consulting Services
		5417	Scientific Research and Development Services
31	Social Services	624	Social Assistance
32	Telecommunications	517	Telecommunications
33	Transportation & Pipelines	48	Transportation
		4921	Couriers
		4922	Local Messengers and Local Delivery
34	Utilities	221	Utilities
35	Warehousing	4931	Warehousing and Storage
36	Wholesale	42	Wholesale Trade
37	Management of Companies	55	Management of Companies and Enterprises

Classification of Industries by Risk Level

Industry sectors were classified into four risk groupings as defined by BJS and DHS. Risk levels were defined as critical, high, moderate, and low. These levels are based largely on whether the industry is part of the U.S. critical infrastructure, as defined by the Information Sharing and Analysis Centers (ISACs). Table 2.3 shows which specific industries were assigned to each level of risk.

Table 2.3
Risk Level and Assignment of Industry Sectors

Risk Level	Industry of Analysis
CRITICAL	5.AGRICULTURE
	9.CHEMICAL & DRUG MFG
	11.COMPUTER SYSTEMS DESIGN
	13.FINANCE
	16.HEALTH CARE
	18.ISPs & WEB SEARCH PORTALS
	24.PETROLEUM MINING & MFG
	26.PUBLICATIONS & BROADCASTING
	27.REAL ESTATE
	32.TELECOMMUNICATIONS
	33.TRANSPORTATION & PIPELINES
	34.UTILITIES
HIGH	20.MFG DURABLE
	21.MFG NON-DURABLE
	25.PICTURE & SOUND RECORDING
	29.RETAIL
	30.SCIENTIFIC RESEARCH & DEVELOPMENT
	36.WHOLESALE
MODERATE	2.ACCOUNTING & BOOKKEEPING
	4.ADVERTISING & OTHER TECHNICAL SVCS
	6.ARCHITECTURE, ENGINEERING & SPECIALIZED DESIGN SVCS
	8.BUSINESS & TECHNICAL SCHOOLS
	17.INSURANCE
	19.LEGAL SERVICES

Table 2.3—Continued

Risk Level	Industry of Analysis
LOW	1.ACCOMODATIONS
	3.ADMINISTRATIVE SUPPORT
	7.ARTS & ENTERTAINMENT
	12.CONSTRUCTION
	14.FOOD SERVICES
	15.FORESTRY, FISHING & HUNTING
	22.MINING
	23.OTHER SERVICES
	28.RENTAL SVCS
	31.SOCIAL SERVICES
	35.WAREHOUSING
	37.MANAGEMENT OF COMPANIES

Cleaning of the Sampling Frame and Sample

Once the fielding period ended, we proceeded with the cleaning of the sample and the sampling frame. This step is necessary to compute the actual sampling probabilities (or sampling weights).

The cleaning of the sample and of the sampling frame was done in two steps. When we acquired the company records from D&B, we generated several sample allocations with different sample sizes. In particular, we generated one with a sample size of 40,000 even though our target sample size was 36,000. The reason for doing so was to ensure that, after cleaning, we would have indeed a final sample size of 36,000.

Step 1: Some of the 40,000 records purchased from D&B were found either to be out of business or to have merged with other companies (n=998). An additional 98 companies were found to be ineligible, and another 70 companies were judged to be duplicate listings. For every stratum, we computed the percentage of companies that were found ineligible in the 40,000 sample. We then used these percentages to clean the frame (i.e., we subtracted those percentages from the initial number of companies in every stratum). This step was completed prior to the fielding period.

Step 2: During the fielding, some of the companies in the sample were found to be ineligible (n=117) or out of business (n=294). Similarly to step 1, for every stratum, we computed the overall rate of ineligible and out-of-business companies. We then used these rates to clean both the sample and the sampling frame. After this cleaning, the actual sample size was 35,590.

After we cleaned the sampling frame (with the information collected during the fielding period), the estimated population size was 7,278,109.

CHAPTER THREE

Sampling Strategy

Overview of the Stratified Sampling Strategy

For the NCSS, we adopted a stratified sampling design, since an important objective of the survey was to provide accurate estimates not only at the national level but also at the industry and company-size levels.

The sampling unit was the company, whether stand-alone or part of a large conglomerate. Branches and divisions were not considered a company; however, subsidiaries within large corporate trees were.

The variables we used to define the strata were industry (a 36-level categorical variable) and company size (a nine-level categorical variable). The U.S. economy was divided into 36 industries. For the definition of the 36 industries, we refer the reader to Table 2.2 in Chapter Two. Company size was defined as the total number of employees of a company. The total number of employees was defined as the sum of the employees at all the branches, divisions and subsidiaries belonging to a company in the case of a company with a complex corporate structure. This variable was then made categorical with nine size classes.

The strata were, therefore, defined by the interaction of these two variables. We also added nine other strata, one for every employment size class, called certainty strata. These nine certainty strata contain companies from the 36 industries that, because of their economic importance or size, were included in the sample with probability 1. So in total, we have 333 strata; however, 81 of the strata were empty, resulting in a total number of 252 sampling strata with at least one eligible sampling unit. The primary reason that some of the strata are empty is the certainty companies (for the definitions of certainty companies, please see the next section, Certainty Companies). In fact, in some instances, there is complete overlap between the larger companies and the certainty companies, therefore leaving many sampling strata empty.

The reason for creating the nine certainty strata was to make the sampling allocation cleaner. That is, in this way, companies sampled from one given stratum are sampled with the same sampling probability. If we had left the certainty companies in their own stratum, defined by their industry and company size, we would have ended up sampling some companies with probability 1 and some with a probability less than 1 for some of the strata.

Certainty Companies

As noted, we sampled with 100-percent certainty several groups of companies because of their importance to the U.S. economy. We refer to these companies as certainty companies and refer to the process for including them as certainty sampling. In agreement with BJS, we decided that the companies satisfying the following criteria would be included in the sample with probability 1.

Criteria defining the certainty companies were:

- Fortune 1,000 companies, including all subsidiaries with 1,000 or more employees
- 50 largest companies (by employment size) within each industry sector
- companies with 5,000+ employees.

Treatment of Subsidiaries of Certainty and Noncertainty Companies

As discussed in Chapter Two, for the NCSS, the company was defined as the sampling unit. This meant that subsidiaries that were part of a large corporation but are companies in their own right were considered eligible sampling units. In particular, it was decided to treat companies (e.g., subsidiaries) belonging to some other company as independent sampling units. In other words, if a company owned another company, that link was broken, and both companies independently had a positive probability to be part of the sample (or the presence of a subsidiary in the sample did not depend on the inclusion of its parent company in the sample).

The subsidiaries of some of the certainty companies were treated slightly differently only because of their overall importance. Criterion 1 in the list above states that not only the Fortune 1,000 ultimates (ultimate is the apex company of a large corporate tree) were included in the sample with probability 1, but also all their subsidiaries with 1,000 or more employees. The subsidiaries of the Fortune 1,000 companies with fewer than 1,000 employees were treated as any other subsidiary instead. That is, they had a positive probability of being included in the sample, and their inclusion in the sample did not depend on the inclusion of their ultimate.

In summary, all subsidiaries, with the exception of the subsidiaries of the Fortune 1,000 ultimates with 1,000 or more employees, had a positive probability to be sampled depending on the stratum to which they belonged.

Sample Allocation

As mentioned earlier, the NCSS data were collected with the intention of addressing multiple objectives. For example, the NCSS data were intended to provide accurate estimates of computer incidents by industry, company size, and whether they occurred at critical-infrastructure companies. It was therefore necessary to find a sample allocation that would be a good compromise in achieving these contrasting goals. In fact, the optimal sample allocation that maximizes precision across industries might not give adequate precision across company-size categories. For this reason, we considered several allocations and chose the one that provided more homogenous effective sample sizes (ESSs) and precisions across industries and company-size classes.[1] The adopted allocation was an adaptive one.

---

[1]   In this case, the ESS is the number of observations from a simple random sample needed to obtain an estimate with sampling variation equal to the sampling variation obtained when accounting for the finite population.

Given a sample size of n sampling units to be allocated, we first allocate n_c sampling units to the certainty strata. In other words, n_c is the total number of certainty companies. We then allocate the remainder sample size $(n-n_c)$ in the following way.

1. Allocate one sampling unit to all the nonempty strata (note that the certainty strata are now excluded, since the certainty companies have already been sampled or allocated). If k is the number of nonempty strata, then k sampling units are allocated.
2. Compute the ESS for every stratum, and assign the next sampling unit to the stratum with the smallest ESS.
3. Repeat 2 until $(n-n_c-k)$ sampling units have been allocated.

The ESS for stratum i is computed in the following way: $\text{ESS}_i = n_i \times (N_i-1) / (N_i-n_i)$, where n_i is the number of observations allocated to stratum i and N_i is the total number of companies in that stratum. Note that the ESS_i is just given by the product of the sample size n_i for stratum i and the finite population correction for that stratum, since, within a stratum, every company has the same probability of being sampled.

This allocation algorithm has several advantages. In particular, it can be easily modified to take into account other characteristics of interest. The algorithm as described deems all the strata equally important; however, it is easy to incorporate weights that weigh some industries more than others or give more weight to larger companies, for example. Lastly, the number of observations allocated to every stratum is an integer, and, therefore, a sample size of exactly of n units can be allocated.

We considered seven different sample allocations. One allocated the sampling units proportionally to the size of the strata. All the others used the algorithm just described. However, two allocations gave more weight to the larger companies. Two other allocations gave more weight to industries with a high risk rating. One allocation was a hybrid of the previous two types; i.e., it gave more weight to larger companies and to companies in higher-risk industries. Finally, the last allocation was the one described in the algorithm just given. We then assessed the seven allocations computing several summary statistics, such as the overall design effect, a weighted design effect with importance weights defined by company size and ESS, and standard errors by industry and company-size class. The chosen allocation had a very small weighted design effect and tended to have more-homogenous ESS and standard errors across industries and company-size classes than all the other considered allocations, therefore representing the best compromise for the type of objectives that this survey data were intended to address.

Frame and Sample Summary Statistics

Table 3.1 presents a summary of the sampling frame and summary statistics for the sample itself and respondents. The table is stratified by risk level and industry sector.

Table 3.1
Frame and Sample Summary Statistics

Industry Number and Risk Level	Industry Name	Universe		Sample			Respondents		
		Companies	Employment	Certainty	Noncertainty	Percent of Industry	Number	Percent of Sample	Percent of Industry
Total		7,278,109	125,032,253	4,822	30,774	37.8	8,189	23.0	10.1
Critical		1,680,606	33,589,833	1,587	10,107	41.6	2,763	23.6	13.2
5	Agriculture	179,442	1,028,733	46	694	21.0	176	23.8	7.1
9	Chemical & Drug Mfg	15,998	2,244,994	183	869	80.0	202	19.2	24.2
11	Computer Systems Design	80,289	1,666,316	57	875	42.3	172	18.5	10.4
13	Finance	187,044	4,570,599	257	1,018	51.8	329	25.8	14.4
16	Health Care	577,499	12,170,633	336	1,108	22.9	426	29.5	10.0
18	ISPs & Web Search Portals	23,874	767,894	60	716	69.8	135	17.4	13.6
24	Petroleum Mining	5,247	374,347	58	509	91.0	132	23.5	12.2
26	Publications and Broadcasting	63,758	2,181,880	139	947	50.7	226	20.8	8.8
27	Real Estate	352,625	2,423,480	56	893	18.0	175	18.4	5.7
32	Telecommunications	26,547	1,159,837	78	743	79.9	136	16.6	9.4
33	Transportation and Pipelines	156,433	4,191,212	157	989	51.4	309	27.0	21.5
34	Utilities	11,850	809,907	160	746	75.1	345	38.1	33.9
High		2,074,041	46,621,994	1,724	5,840	44.5	1,744	23.5	11.0
20	Mfg - Durable	275,319	14,890,589	737	1,122	48.4	519	27.9	14.1
21	Mfg- Non-Durable	127,248	6,193,560	304	1,067	46.5	334	24.4	13.6

Table 3.1—Continued

Industry Number and Risk Level	Industry Name	Universe		Sample			Respondents		
		Companies	Employment	Certainty	Noncertainty	Percent of Industry	Number	Percent of Sample	Percent of Industry
25	Picture and Sound Recording	31,902	540,558	47	595	64.7	89	13.9	32.6
29	Retail	1,005,175	15,945,636	378	1,040	51.6	322	22.7	9.1
30	Scientific Research and Development	185,571	2,569,481	88	965	29.0	221	21.0	7.2
36	Wholesale	448,825	6,482,170	170	1,051	21.0	289	23.7	6.1
Moderate		774,494	8,464,645	552	4,742	37.0	1,196	22.6	8.3
2	Accounting and Bookkeeping	96,023	788,299	48	627	39.6	177	26.2	6.0
4	Advertising and Other Technical Services	121,116	1,389,991	65	811	36.9	154	17.6	6.4
6	Architecture, Engineering, and Specialized Design	131,291	1,508,863	57	840	24.9	217	24.2	6.7
8	Business and Technical Schools	51,275	605,531	46	710	28.6	190	25.1	9.3
17	Insurance	174,108	2,794,075	285	958	54.7	276	22.2	12.0
19	Legal Services	200,680	1,377,885	51	796	16.6	182	21.5	5.4
Low		2,748,969	36,355,783	959	10,085	25.9	2,456	22.2	6.5
1	Accommodations	60,944	1,981,664	84	922	47.5	143	14.2	4.4
3	Administrative Support	355,532	6,869,345	242	1,055	33.2	256	19.7	6.9
7	Arts and Entertainment	145,913	2,087,732	59	898	25.0	199	20.8	5.6
12	Construction	692,084	6,932,918	93	1,010	8.7	245	22.2	3.4
14	Food Service	277,281	7,453,100	114	1,015	36.5	213	18.9	9.9

Table 3.1—Continued

Industry Number and Risk Level	Industry Name	Universe				Sample			Respondents		
		Companies	Employment	Certainty	Noncertainty	Percent of Industry	Number	Percent of Sample	Percent of Industry		
15	Forestry, Fishing, and Hunting	29,132	270,348	49	583	29.5	153	24.2	7.0		
22	Mining	15,082	562,764	65	730	66.5	195	24.5	28.4		
23	Other Services	897,994	6,446,578	57	982	12.0	273	26.3	4.0		
28	Rental Services	62,970	872,965	52	741	47.1	161	20.3	8.2		
31	Social Services	180,376	2,319,343	49	918	15.4	317	32.8	6.6		
35	Warehousing	18,701	224,628	47	557	46.5	132	21.9	9.3		
37	Management of Companies	12,960	334,396	48	674	71.5	169	23.4	10.0		

CHAPTER FOUR

Survey Methods and Fielding

Prior to RAND contracting with BJS to field the NCSS, BJS did a great deal of preparatory work in conjunction with the U.S. Census Bureau to develop a pilot computer-security survey instrument (at that time, called CSS) and to test the feasibility of fielding such a study among U.S. businesses. In 2001, the Census Bureau administered the CSS to 500 companies as a pilot test, whose "findings . . . illustrate[d] the feasibility and utility of a data collection program to be initiated in 2004 among some 36,000 businesses. . . .Results of this test demonstrated a need for an increased response rate to produce valid national estimates and a need to refine survey questions" (Rantala, 2004, p. 1). BJS subsequently awarded a contract to RAND in the fall of 2004 to field the national survey, designed to be nationally representative of U.S. businesses' experiences with computer-security incidents. The RAND project team began work shortly thereafter on the instrument adjustments and fielding design and methodology needed prior to implementation.

This chapter documents the overall testing of and revisions to the NCSS instrument with some comparisons to the 2001 CSS instrument; the initial fielding-strategy design and testing; fielding-strategy revisions based on testing; fielding implementation; and a summary of final response rates.

Instrument Development

The 2005 NCSS developed by BJS and given to RAND for testing was based closely on the 2001 CSS instrument used in the pilot test, with slight modifications made given changes in technology over a five-year period and with the addition of a limited number of new questions. BJS asked that RAND team–recommended changes to the instrument steer clear of a wholesale rewrite that might diminish the benefit of the question-and-response–set testing that the Census Bureau had done prior to the fielding of the 2001 pilot.

In the Office of Management and Budget (OMB) application, three versions of the survey instrument were proposed, each tailored to a specific company size to reduce respondent burden. There were three versions that RAND tested:

- standard form, for first and second packet mailings to companies with 25+ employees
- short form, for small companies (with fewer than 25 employees)
- conversion form, for those receiving the long form initially, to be used for nonrespondent conversion via the third and fourth survey-packet mailings.

RAND's Survey Research Group (SRG) tested the instruments by carrying out in-depth cognitive interviews. The cognitive-interview task was designed to achieve two goals: (1) identify response error in the NCSS so as to maximize the reliability and the validity of the survey data gained through the NCSS instruments and (2) obtain information that would help inform the endorsement strategy for the survey. A third unintended, yet important, result of the interviews was the capture of information on company information-technology (IT) structure, which later became critical to informing the NCSS sampling design and amending the survey instructions for companies regarding which parts of their companies they should report on (discussed in "Company-Structure and Fielding-Strategy Considerations").

The RAND project team conducted 27 cognitive interviews with IT executives and related staff from a diverse range of industries and company sizes across the nation. Three separate protocols were used for this effort. The first protocol was based on the short version of the questionnaire and concentrated on testing both the response process and comprehension of the specific questions and response items. The second protocol represented the standard version of the questionnaire and focused on testing the overall flow of the questionnaire, concentrating on comprehension of the questions, the terms used, and the respondent burden involved. The third version of the protocol was based on the details of Section III, "Types of Computer Security Incidents" in the questionnaire. Specifically, Section III asks respondents a series of questions about their experience with each of specific types of computer-security incidents, including embezzlement, theft of IP, fraud, computer viruses, electronic vandalism or sabotage, denial of service, and other computer-security incidents. The third version of the protocol tested both memory recall and definitions of the different terms used. Each interviewee was assigned to one of the three protocols, so that each interview focused exclusively on the items in that selected protocol.

Probes used in these semistructured interviews were based on items about which the BJS Project Manager had expressed concern, as well as items that the RAND project team anticipated could be problematic. Overall, testing priority was given to new questions, questions with new response categories, questions about computer-security incidents, questions about monetary loss and costs, and questions related to networks accessed or affected, routes, reporting of incidents to organizations, reasons for not reporting, and offender/company relationship.

Key Modifications Made to the Survey Instrument

Overall, several major themes emerged, which fell generally into the following categories: clarity regarding improved ordering of question sets; individual questions deemed to be repeatedly problematic (linked to findings that the one-size-fits-all approach to question construction was not always successful); concerns about the survey length; and the importance of featured private-sector support and readily available benchmark reports as incentives to completion. The RAND team and BJS then met to discuss the recommendations, BJS approved a certain number of these and requested a few additional changes, and a final survey instrument was agreed upon. The macro-level changes recommended by the RAND team and accepted by BJS are documented next.

Restructuring of Survey Order. Cognitive testing revealed that certain panel sections were better suited to the technology-oriented audience being targeted, making these sections easier to answer for respondents and fostering a more positive (and less frustrating) initial experience with the survey. These sections (computer viruses, denial-of-service attacks, electronic vandalism or sabotage) were more regularly encountered by the cognitive-testing respondents, who, in turn, reflected the target audience for the 2005 NCSS. The testing team found that the remaining incident sections (embezzlement, fraud, and, to a lesser degree, theft of IP) were likely to need input or answering in full by other parties in the organization (and also were considered by the cognitive-interview subjects to be more sensitive in nature), strongly arguing for their inclusion following the sections that the targeted respondents could more easily answer. This reordering was tested during the latter half of the cognitive interviews and found to flow more naturally with respondents. Given this, the RAND team recommended reordering of Section III for the panel-question sets accordingly. This recommendation was accepted by BJS and implemented in the 2005 NCSS.

Modification of Question-and-Response–Set Wording. Additionally, changes to some of the questions and response sets were recommended for minor modification based on interviewee feedback to improve data capture. BJS accepted some of these recommendations (e.g., divide current employees and current "other" workers into two separate perpetrator-response categories, as interviewees perceived them to contain substantially different risks), while others were not (e.g., collapse all "hacker" response categories into one with the option with an open comment field for recording country of origin, if known). The RAND team also recommended the addition of a question regarding computer-security incidents resulting from information obtained from a stolen laptop computer, given the rise in incidents based on this access point. BJS agreed to this addition. Finally, BJS requested that the 2001 section called Theft of Proprietary Information be divided into two sections: Theft of Intellectual Property and Theft of Personal or Financial Information. This change was executed as requested.

Company Structure. One of the final revelations from the cognitive interviews was that an assumption made earlier on—that companies' IT structures for most conglomerates were centralized—was found not to be true in approximately half of the large companies interviewed. Instead, we found wide heterogeneity in how IT was structured among complex companies. For some, IT was integrated top to bottom, and a single chief security or technology officer at the ultimate company level could answer the survey's questions about incident experiences across companies within that corporate family. For others, each company had its own IT structure, and, for an accurate report, each company in that family would have to be contacted. Still other corporate families reflected a mix of the two, with part of the corporate family integrated from an IT perspective but with other more recently acquired companies still operating and reporting IT activities separately. This limited some ultimate companies' ready access to information about their subsidiaries' experiences with computer-security incidents and, consequently, the ultimates' ability to report for all their subsidiaries.

The information from the cognitive interviews also informed our sampling design and fielding protocol (to be discussed in greater detail in "Fielding Considerations"). The survey instrument's instructions were revised to ask each company to report for all of its divisions and locations but to exclude subsidiaries, which could be contacted separately if sampled. However, knowing that the separation of this information would not be possible for all companies, a final question was added to the survey instrument to capture for whom the survey had been completed: "Does the information reported in this survey include this company or does it include this company and some or all of its subsidiaries?" If the response was SOME or ALL, the survey asked, "How many subsidiaries were included?" and "Please list subsidiaries included in this report." This was a departure from the 2001 survey instrument, which asked companies to report for all divisions, locations, and subsidiaries. Appendix B provides the final questionnaire for the NCSS.

Fielding-Strategy Considerations

Subsidiaries

As discussed previously, a key question for the project team during the design phase was how to most effectively survey large, complex companies with multiple subsidiaries. Two decisions made early on in the project affected the level of complexity faced during fielding:

- As stated in the request for proposals (RFP), the NCSS would be a company-level versus an establishment-level survey.
- Each company and its subsidiaries would be placed in its proper industry sector to permit cross-sector comparisons.

To do justice to both of these objectives would require collecting information not only from each ultimate but also from a number of its subsidiaries separately. This meant that, for the large conglomerates, a large portion of the sample would potentially be tied up surveying these companies' corporate family trees.

Following the 2001 pilot survey strategy of sending one survey per company family to each these large conglomerates could lead to several problems:

- The respondent at the ultimate level of the pyramid may not have information for all the entities below it or may have this information for some types of questions (embezzlement, fraud) but not for others (denial-of-service attacks, computer-virus attacks).
- Analysts of the data would not know exactly for which entities within the organizational pyramid the respondent had reported.
- Companies within a pyramid could belong to multiple industry sectors that differ from that of the ultimate, making a clean breakdown of the data by industry sector impossible.

Thus, for the large conglomerates, a decision had to be made about which entities to treat as the reporting unit and a tailored survey-fielding and analytic strategy devised to ensure that

- the data we received were as complete as possible
- it was clear for which entities such data had been reported
- there was a strategy in place for dealing with omitted entities (i.e., when the ultimate responded only for itself but not its subsidiaries)
- at the time of analyzing the data, there would be agreement about how the large conglomerates would be handled.

The decision was made to relax the definition of *company*, instead treating each subsidiary headquarters as both a sampling and reporting unit. We elected to survey the ultimate and a sample of subsidiaries beneath it individually. Each unit surveyed (the ultimate and the sample of subsidiaries beneath it) was asked to report information only about itself. As aforementioned, this required modifications to the survey instructions about the definition of *reporting entity* and adding several new questions to the survey for data-validation purposes.

This decision also had implications for the survey-fielding strategy and analytic strategy regarding the reconciliation of overlapping cases. SRG staff allocated more time to reconcile inquiries from respondents about whether a company should report for itself as well as its subsidiaries and what to do if a respondent were unable to disentangle information about itself from its subsidiaries. In many instances, inquiries were treated on a case-by-case basis, and, throughout the field period, new status codes were assigned to account for subsidiaries included with parent-company responses or parent companies responding for nonsampled subsidiaries.

Confidentiality

The need for confidentiality on the NCSS was of the utmost importance, since businesses were asked to provide sensitive information about computer-security issues that, if made public, could be detrimental to their reputations and the public's perception of them. The agreement into which RAND entered with BJS and federal statutes under which the data were collected required that confidentiality be protected. The strategy for adhering to data-confidentiality and privacy-certification legislation included several parts:

- the development of a detailed data-safeguarding plan (DSP),[1] reviewed and approved by RAND's Human Subjects Protection Committee (HSPC)
- a security audit of the vendor hosting the Web-based version of the survey
- a data-safeguarding review of the print vendor.

RAND's DSP specified the standard operating procedures for the transmission of project data and their storage on RAND computer systems. Electronic transmission of data required the use of encrypted files using PGP-software[2] public/private key combinations. In other instances, removable media were used to move data from one computer to another. The manual transfer of data was required, as the principal analyst used a computer that had no network access. Where collaboration was required, storage of computerized files on networked computers (e.g., the SRG phone center) was always on appliances in secured locations using password-protected computers that had no Internet access. In all instances, the sample data containing explicit identifiers were separate from substantive survey data.

Symantec was selected to conduct an assessment of RAND's internal and Internet presence. This included the detailed security audit of the Web-hosting vendor, Market Strategies International (MSI) and was expanded to include an application assessment of RAND's Internet survey application and its host server. The audit of MSI was based on the ISO 17799 standard revised for this project's requirements and including an application assessment of the Internet survey software employed. Beyond MSI's overall security posture, the project instituted procedures designed to "blind" MSI to respondents' identities—both as companies and individuals.

Language regarding applicable privacy legislation was included on the survey instruments and the Web site. The survey implementation did not experience any breaches in data confidentiality. While a small number of companies self-identified via open-ended responses in the survey, confidentiality was maintained through several efforts: (1) in the analytic file, only Case IDs appear, not any company-identifying information and (2) all identifying information provided in open-ended questions was cleaned of identifying information before being

---

[1] The DSP outlines overall responsibility for data safeguarding, how respondent identity will be protected, procedures for data transmittal, disclosure risks, audit and monitoring plans, client and respondent agreements, and data-safeguarding procedures.

[2] PGP stands for *pretty good privacy*.

delivered to BJS. In this way, the data file would not reveal specific responses of a particular company that may have self-identified.

The project team periodically reviewed the DSP with RAND's HSPC data-safeguarding expert and amended it to address unanticipated issues that arose during the course of fielding the survey.

Additional data-safeguarding efforts were undertaken in relation to the print vendor. The vendor had extensive experience in handling sensitive information for national educational and health surveys and so already was familiar with privacy-protection procedures. To ensure that the data-protection requirements were met for the NCSS, the RAND team additionally visited the printing facility to review its procedures for processing and handling materials and compliance with data-protection standards; required that the print vendor and subcontractors sign a nondisclosure agreement; and ensured that the vendor's computing systems were on isolated networks with no Internet access.

Endorsements

The RAND team and BJS worked hard to gain trade associations' endorsements. The RAND team met either in-person or by telephone with more than 35 trade associations, participated in conference calls and individual trade-association meetings, and conducted telephone and email follow-up. Associations were asked to help get the word out about the survey by posting announcements in their newsletters or on their organizations' Web sites, to provide quotes in support of the survey for use on the project Web site, and to send messages to their memberships to encourage participation in the survey. In addition to following up with previous contacts, BJS also called and met with more than a dozen organizations. Collectively, the RAND team and BJS had contacted approximately 50 organizations. A total of 32 organizations agreed to be listed as project supporters (Table 4.1). In addition, quotes for use on the project Web site were obtained from 15 organizational leaders.

The RAND team developed a project Web site that listed the names of project supporters and informed industry, survey respondents, and endorsers about the purpose of the survey, provided additional information about the study, and encouraged participation. The Web site included a description of the project, information about the sponsors (DOJ/BJS and DHS/National Cyber Security Division [NCSD]), contact information for RAND and BJS, NCSS supporters' information (includes a list of trade associations who have agreed to support this effort and quotes of support from trade-association leaders), frequently asked questions (FAQs), respondent information about how to participate in the survey, and privacy-policy information.

Table 4.1
Final List of NCSS Project Supporters

American Association for Technology in Psychiatry

American Insurance Association (AIA)

American Chemistry Council (ACC)

American Society for Clinical Pathology (ASCP)

ASIS International

Association of American Railroads

BITS, The Financial Services Roundtable

Business Executives for National Security (BENS)

Business Software Alliance (BSA)

Computer Emergency Response Team Coordination Center (CERT/CC)

Cyber Security Industry Alliance (CSIA)

Electric Power Supply Association (EPSA)

Food and Agriculture Information Sharing and Analysis Center (Food and Ag ISAC)

Healthcare Information and Management Systems Society (HIMSS)

Information Technology Association of America (ITAA)

Information Technology–Information Sharing and Analysis Center (IT-ISAC)

InfraGard

Joint Council on Information Age Crime (JCIAC)

Manufacturers Alliance/MAPI

Medical Group Management Association (MGMA)

National Alliance for Health Information Technology (NAHIT)

National Association of Manufacturers (NAM)

National Coordination Office for Networking and Information Technology Research and Development (NCO/NITRD)

National Federation of Independent Businesses (NFIB)

National Petrochemical and Refiners Association (NPRA)

National Telecommunications and Information Administration (NTIA)

President's Council of Advisors on Science and Technology (PCAST)

The Real Estate Roundtable (RER)

Risk and Insurance Management Society, Inc. (RIMS)

Small Business Group and Entrepreneurship Council (SBEC)

U.S. Chamber of Commerce

Women Entrepreneurs, Inc. (WE Inc)

Multiple Fielding Waves

The NCSS was fielded in two waves, referred to as Wave I and Wave II. Wave I was designed to test various modes and fielding methods across company-size categories to inform expectations and fine-tune Wave II fielding efforts. For this test, we created five treatment groups (see Table 4.2) based on survey length (long versus short), multi- versus single-PIN and Web versus paper. To accentuate the potential difference in mode effects between companies that would receive packets containing a paper copy of the instrument and those that would not, we determined that the contents of the second packet would mirror what a company received in the first packet, with everyone in the sample receiving a paper survey with the third packet mailing, regardless of previous treatment assignment. (Additionally, due to a request made by the American Chemistry Council that had an industry survey being fielded in late spring, the entire chemical sector was fielded in Wave I. The treatment protocols for this group were

Table 4.2
Wave I Modal Test Design

Group	n	Description	Survey Packet Mailing			FedEx
			First	Second	Third	
A	300	Long multi-PIN Web	No paper survey	No paper survey	Conversion survey	Conversion survey, selected companies
B	900	Long single-PIN Web	No paper survey	No paper survey	Conversion survey for large companies and half of the medium-sized companies; short survey for the remaining half of the medium-sized companies	Conversion survey to selected large companies
C	1,000	Long single-PIN paper	Long survey	Long survey	Conversion survey for large companies and half of the medium-sized companies; short survey for remaining half of the medium-sized companies	Conversion survey to selected large companies
D	300	Short Web	No paper survey	No paper survey	Short survey	NA
E	300	Short paper	Short survey	Short survey	Short survey	NA
F	250	Long multi-PIN Web, chemical	No paper survey	Long survey	Conversion survey	Conversion survey to selected companies
G	325	Long single-PIN Web, chemical	No paper survey	Long survey	Conversion survey	Conversion survey to selected companies
H	97	Short Web, chemical	No paper survey	Short survey	Short survey	NA

identical to those expected to be used in Wave II [becoming groups F–H], creating eight treatment groups in all.)

Findings of the test included the following:

- **Differences in response rate between the slightly longer conversion instrument and the short instrument, when issued in the final packet mailing, as a conversion tool.** *Outcome:* Following the design of the Wave I test but prior to the analysis of the differences in response between the conversion and short instruments, additional consideration was given to the importance of being able to produce national- and industry-level estimates on some types of loss estimates based on questions not available on the short form, versus a lower response rate. As a result, the decision was made to forgo all use of the short form during the Wave II fielding, sending the conversion instrument to small companies from the outset instead. It was determined that the benefits of being able to produce these estimates outweighed the risks of lowering response rates for four reasons:
  - There were few financial questions on the conversion form.
  - Those questions did not appear until the middle of the form.
  - The respondent would need to answer only those questions that pertained to the types of incidents the company had experienced.
  - Smaller companies were expected to have few different types of incidents.
- **Use of the multi-PIN option by large companies, given that the 2001 pilot and our cognitive interviews indicated that companies would like the capability to fill out different sections of the survey.** Specifically, large companies were provided with a master PIN to access their Web surveys, as well as four additional PINs to allow for distinct users to log in only to specific sections of the survey.

The survey sections were grouped as shown in Table 4.3 for the multi-PIN option:

Table 4.3
Survey Sections for Multi-PIN Surveys

Test Portion	Topics
Sections I, II, IV, and V	Computer Security Concerns, Computer Infrastructure & Security, Other Trends, and Company Information
Section III, questions 5–7	Computer Virus, Denial of Service, and Electronic Vandalism or Sabotage
Section III, questions 8–11	Embezzlement, Fraud, Theft of Intellectual Property, and Theft of Personal or Financial Information
Section III, question 12	Other Computer Security Incidents

Outcome: Although in Wave I only a handful of companies issued the multi-PIN option used it instead of the master PIN, there did not appear to be a detrimental effect on response in offering the multi-PIN versus the single-PIN option for companies of similar size.

- **Response-rate differences between companies receiving packets initially containing a paper version of the instrument and information on accessing the Web version of the survey versus those that received packets containing information only about how to access the Web version.** *Outcome:* Initial response from companies receiving packets containing a paper survey was greater than that from companies receiving initial packets with information only about accessing the survey via the Web. However, given the significant costs associated with a strategy change that would involve printing and mailing paper surveys to the entire sample as part of the initial packet, along with an executive decision by BJS to maintain the original sample size of 36,000, mailing paper surveys to all companies was not an option, given cost constraints. Thus, the initial packet-mailing strategy (outlined next) was maintained for Wave II.

Multiple Modes

It has been widely accepted that fielding multiple modes of a survey and attempting multiple rounds of follow-up contact serve to increase survey-response rates (Dillman, 2007). Given this, and to help contain the cost of surveying 36,000 businesses, the NCSS was fielded initially as a Web-based survey, followed by an option to complete the survey in a paper-and-pencil format. Further, a fixed, or nonfillable, portable document format (PDF) file was designed so that each company in the sample could access a PDF copy of the survey online, each preloaded with that company's corresponding unique identifier, thus mimicking the paper version of the survey. This version could be either faxed or mailed back to RAND once completed. The fixed, downloadable PDF was made available to all respondents from the start of the first mailing, from the same password-protected Web site as the Web version.

Taking all of this into account, the following overall fielding strategy and plan for nonresponse follow-up was designed:

1. precall (Wave I) or Web lookup (Wave II) to research and update company addresses prior to packet mailings (outcomes discussed next)
2. initial invitation packet with personalized letter from the RAND project leader, Web URL with PIN information and list of industry-related project supporters
3. two weeks later, a reminder letter in a sealed envelope
4. three weeks after first packet, a second packet containing a preaddressed, postage-paid return envelope and a hard-copy survey for nonrespondents
5. five weeks after first packet, a third packet containing a return envelope and hard-copy survey for nonrespondents
6. follow-up phone calls to a portion[3] of the nonrespondent sample
7. six weeks after first packet, final hard-copy survey via FedEx to a portion[4] of the non-respondent sample.

Precalls and Web Lookups

The Wave I pretest allowed us to analyze results from our precall efforts to companies. The original purpose of precalling in Wave I was twofold:

-   Confirm that we had the most appropriate contact person on record for the company.

---

[3] Follow-up phone calls were made to the 10 percent of cases deemed the most important in each wave. For Wave I, follow-up calls were made to all Fortune 1000 companies and all certainty companies in the critical risk category (n = 331). For Wave II, follow-up calls were made to all certainty companies in the critical and high risk categories (n = 3,272).

[4] We planned to FedEx hard-copy surveys to the 10 percent of cases deemed most important in each wave. For Wave I, FedEx packets went to all certainty companies in the critical and high risk categories (n = 792). For Wave II, FedEx packets went to all certainty companies across all risk categories (n = 2,602).

- Confirm or update the mailing address in the database as the best mailing address for that contact and company.

The Wave I precalling effort made apparent that attempting to confirm or update the best contact person for the survey was a challenging task in advance of the survey mailing. Because the companies had not yet received any project materials and got only a very brief overview of the survey over the phone, the task of asking them to confirm the name of the appropriate contact person was complex and far more extensive than precalling activities would normally entail. As a result, the precalls ultimately focused on confirming and updating company mailing addresses. However, even obtaining this simple information for a number of cases took several rounds of routing through company phone systems, along with repeated explanations of the purpose of the mailing information. At the end of the Wave I precalling effort, 2 percent had confirmed their company's name and address; 29 percent updated their name or address; and 2.5 percent were flagged for supervisor review (n = 853).

Since the Wave I precalls yielded only very basic information (mailing address) and even then were relatively inefficient (with an average of about 12 minutes per case), we switched to a Web lookup effort in place of the precalls to confirm and update mailing addresses for Wave II. Using this method, we saw the following improved results: 67 percent of the company names and addresses were confirmed through the Web lookup process, 29 percent were updated, and 4 percent were flagged for supervisor review (n = 3,951).

Incentive

All sampled companies were promised that they would be provided with a benchmark report for their company's particular industry sector if they participated in the survey. While these benchmark reports have not yet been made available (because the release of the BJS report must precede them), the strategy by which the benchmark reports will be made available to companies follows. All companies that responded to the survey will be mailed a letter thanking them for their participation and directing them to a Web site from which their particular industry sector's benchmark report can be downloaded. Each letter will contain a PIN to allow the company access to the report for its sector. These letters are to be mailed as soon as the BJS summary results for the NCSS are released.

Survey Implementation

Wave I. A total of 3,472 companies received surveys in the Wave I field period. The budget for fielding and printing of materials for Wave I was based on the assumptions that we would send the reminder letter to 100 percent of the sample; that we would have 15-percent response by the time of the second packet mailing; and that we would have 35-percent response by the time of the third packet mailing.

Two weeks following the first mailing, calls into the 800 number and activity on the Web survey began to spike. For a typical nonbusiness survey, this activity would have begun within a few days of the initial invitation being mailed, around the day that packets would arrive via U.S. Postal Service. For this business survey, although packets might have been received by the designated respondent in the same limited number of days for a small business with only a few employees, the routing of the survey packet through corporate mailrooms and into the hands of the designated respondent (and, in some cases, rerouted from there to a more appropriate respondent) took far longer in large businesses with many employees. Consequently, many calls were received at the two-week mark indicating that respondents had just received the survey in their inboxes, some of whom were concerned about the two-week deadline mentioned in the cover letter used for Wave I.

Given this issue, which explained the unexpectedly low response in the first two weeks, and given that two weeks after the initial mailing responses were still slowly increasing rather than decreasing, the RAND team and BJS decided to postpone the second packet mailing by one week while the last round of packets were evidently still "live" in the field. Best-practice survey-research methodology calls for projects to observe a peak and drop-off in response before fielding follow-up packets, and this decision was made with this in mind.

From a fielding-resource perspective and to avoid frustrating survey respondents by continuing with the original fielding schedule, the NCSS fielding strategy was adjusted to allow for more time between mailings. The following revised fielding timeline was executed, with the last FedEx packet mailing taking place eight weeks following the first packet mailing:

1. two weeks after first packet mailed, reminder letter mailed
2. three weeks after first packet, phone follow-up to certainty companies
3. four weeks after first packet, second packet mailed to nonresponders
4. seven weeks after first packet, third packet mailed
5. eight weeks after first packet, fourth packet sent via FedEx to 10 percent of the sample.

Based on the experiences in Wave I, two lessons emerged that informed changes to the fielding strategy for Wave II.

One, as noted, the use of precalls to update address information for companies was not deemed to be the most resource-efficient or effective method for achieving this goal. A Web lookup effort was devised as a replacement to the precalls to confirm or update mailing addresses for Wave II. During testing of the Web lookup task, an attempt was made to find contact names from the company's Web site, but staff found that, like the phone calls, the Internet was not a useful tool for this type of search. Thus, this element of the prefielding updates to records was dropped. Instead, we asked that staff search for each business online and confirm or update the mailing address only. Under this model, the team was able to confirm or update information for more than one case per minute (n=3,951).

Two, important lessons were learned from the Wave I follow-up calls that influenced our planning for Wave II follow-up calls. Most of the lessons influenced elements of interviewer training for Wave II, including informing calling staff that they needed to be particularly prepared to answer questions about confidentiality. Additionally, we learned that, often, contacts had seemingly not received the survey packet because of the complexities of corporate mail routing. This information was helpful for callers, since they needed to be prepared to explain the project to someone who likely had not yet seen any of the materials mailed to them.

In Wave II, changes were also made to the mailing packet. Two letters encouraging participation were added: a cover letter signed by BJS Director Jeffrey Sedgwick and a letter signed by Attorney General Alberto Gonzalez, and Secretary of Homeland Security Michael Chertoff, describing the importance of the survey. And to better identify the official nature of the contents, the phrase "On behalf of the USDOJ and USDHS" was printed on the outside of the mailing envelope.

Wave II. A total of 32,517 cases received surveys in the Wave II fielding. Wave II was executed as listed here:

- Web lookup for address confirmation and update
- first packet mailed, PIN and Web-access information included
- two weeks later, reminder letter mailed, PIN and Web-access information included
- five weeks after first packet mailed, second survey packet mailed, with paper survey included[5]
- six weeks after first packet mailed, phone follow-up (lasted 10 days)
- 10 weeks after first packet mailed, third survey packet mailed, with paper survey included[6]
- 12 weeks after first packet mailed, FedEx packet mailed, with paper survey included.

Download of Web Data. During the field period, data from MSI (the Web vendor hosting the survey) was delivered to RAND through MSI's product known as MSIClient™. Access to MSIClient required a secure certificate issued by MSI with an ID and password. All data files were encrypted using a public key issued by RAND using PGP and exchanged over a secure socket layer. Sample files were delivered to MSI using the same technology with encryption using MSI's public key.

Web Data Retrieval. Toward the end of Wave II fielding period, a programming error was uncovered in the standard Web version of the survey, which resulted in the first financial-loss question value in each computer-incident set (questions 5–12) of Section III being overwritten by the value for the second financial-loss question for each type of computer-security incident. For example, this affected questions where the respondent filled out both Q5j and Q5k and where the respondent left Q5k missing, making it impossible to distinguish between companies that elected not to answer these items and companies whose responses were excluded. This programming error did not affect the Web conversion form, and the short form did not contain these questions. This particular problem was difficult to detect for several reasons: First, the questions in which the problem surfaced were the ones that we expected from the outset to be missing the most data, either because companies did not track this information and, thus, did not have the numbers available or because companies had confidentiality concerns about sharing this information. This meant that, in examining the distribution of missing data, the pattern looked as expected for these financial questions. Second, this programming error did not affect the Web conversion form. And third, when the data were downloaded, they appeared in the affected fields at the same rate as the other financial questions.

Once the source of the error was identified, the RAND team worked with the web vendor to correct the programming error and, after consulting with the HSPC, implemented the fol-

---

[5] As we prepared the second packet mailing, we discovered that 14 percent of the responses that were deemed "good" by BJS were actually Web partials rather than fully completed forms. As a result, second packets were mailed to nonrespondents as well as to Web-survey respondents who were missing at least one complete section (89 percent of the sample). However, second packets were not mailed to Web-survey respondents who answered at least part of each section but may have left some items missing or respondents who returned a paper survey, regardless of its completeness.

[6] Given that the partial cases had already had an opportunity to return to the survey and had not yet done so, and given that the packet mailing and instructions could not be tailored to indicate which portion of their survey was still outstanding, it was felt that the return on mailing this packet to partial responders would be marginal. Thus, the third packet was mailed strictly to nonrespondents (79 percent of the sample).

lowing data-retrieval strategy: the RAND team recontacted by email (or by letter, if the email address was not available) the individuals who filled out the Web long survey form, explaining to them the issue and requesting that they return to an NCSS Web instrument to fill out the information for those few questions. The web vendor posted a revised Web survey that would direct the respondent immediately to only those financial questions in Section III for which they had answered the screener questions affirmatively. Follow up was done by email and by telephone for those who did not respond within two weeks.

In total, 1,391 companies were included in the data-retrieval effort, of which 557 (40.0 percent) responded with some data; 230 companies (16.5 percent) logged into the recontact form on the Web but did not report any data (blank rows); 604 (43.4 percent) of those contacted did not respond.

Fielding Respondent Inquiries. A toll-free number and a project email address were established for respondent questions. The use of both the 800 number and the email alias was higher than expected, and staff time was reallocated throughout the project to cover these tasks. A total of 1,480 incoming calls were received on the 800 number, and a total of 368 emails were received via the project address. The BJS Project Manager also reported receiving more than 130 direct calls.

Many respondents called in to check on the status of their survey or to report updated contact information (394 of 1,480 calls logged, or 26.6 percent). About 21 percent of the incoming calls were related to a request of some kind (e.g., request for a hard-copy version of the survey, confirmation that the survey had been received, need for a misplaced PIN). Approximately the same proportion (20.8 percent) called in about whether the survey was voluntary, what the deadline was for survey response, or similar questions. A total of 273 calls (18.4 percent) were related to issues with the survey (e.g., problems accessing the Web site, a company received multiple surveys). Finally, about 13 percent of incoming calls were to refuse participation in the survey.

Respondents also sent queries to the project's email alias. As with the 800 number, most respondents contacted the project to check on the status of their survey or to report updated contact information (34.5 percent of all emails received). A total of 96 emails (26 percent) were related to issues with the survey. About 16.3 percent of the emails were related to a request of some kind. Emails to refuse participation accounted for 15.7 percent of email traffic, and 7.3 percent of emails were from respondents with questions about the survey.

Data Processing and Record Management. RAND's SRG used a sophisticated record-management system (RMS) to trigger the appropriate follow-up task (reminder mailing versus reminder call). The RMS tracks the current status of each case in the sample, records the date of key events, and acts as inventory control to document the receipt of completed, partially completed, or blank self-administered questionnaires (SAQs). The RMS was used to track both the process status and the outcome of each case throughout the fielding period. Interim statuses were used to inform operations while final statuses were provided to analysts.

Response-Rate Summary

Table 4.4 presents the final response rates by industry and by company size.

Table 4.4
Final Response Rates, by Risk Level, Company Size, and Industry for the National Computer Security Survey

Factor	Number Sampled	Number Responded	Percent Response
Risk Level	35,596	8,189	23.0
Critical	11,694	2,763	23.6
High	7,564	1,774	23.5
Medium	5,294	1,196	22.6
Low	11,044	2,456	22.2
Certainty	4,822	1,429	29.6
Employment	35,596	8,189	23.0
10,000 or more	1,431	407	28.4
5,000 to 9,999	1,353	444	32.8
2,500 to 4,999	4,260	1,090	25.6
1,000 to 2,499	2,991	764	25.5
500 to 999	3,704	976	26.3
100 to 499	4,777	1,206	25.2
25 to 99	5,601	1,245	22.2
10 to 24	5,704	1,165	20.4
2 to 9	5,775	892	15.4
Grand total	35,596	8,189	23.0
Critical	11,694	2,763	23.6
Agriculture	740	176	23.8
Chemical & Drug Mfg	1,052	202	19.2
Computer Systems Design	932	172	18.5
Finance	1,275	329	25.8
Health Care Services	1,444	426	29.5
ISPs & Web Search Portals	776	135	17.4
Petroleum Mining & Mfg	567	132	23.3
Publications & Broadcasting	1,086	226	20.8
Real Estate	949	175	18.4
Telecommunications	821	136	16.6
Transportation	1,146	309	27.0
Utilities	906	345	38.1

Table 4.4—Continued

Factor	Sampled	Responded	Percent Response
High	7,564	1,774	23.5
Mfg Durable	1,859	519	27.9
Mfg Non-Durable	1,371	334	24.4
Picture & Sound Recording	642	89	13.9
Retail	1,418	322	22.7
Scientific R&D	1,053	221	21.0
Wholesale	1,221	289	23.7
Medium	5,294	1,196	22.6
Accounting & Bookkeeping	675	177	26.2
Advertising & OtherTechSvs	876	154	17.6
Architecture, Engineering & Other	897	217	24.2
Business & Tech Schools	756	190	25.1
Insurance	1,243	276	22.2
Legal Services	847	182	21.5
Low	11,044	2,456	22.2
Accommodations	1,006	143	14.2
Administrative Support	1,297	256	19.7
Arts & Entertainment	957	199	20.8
Construction	1,103	245	22.2
Food Services	1,129	213	18.9
Forestry, Fishing, & Hunting	632	153	24.2
Management of Companies	722	169	23.4
Mining	795	195	24.5
Other Svs	1,039	273	26.3
Rental Svs	793	161	20.3
Social Services	967	317	32.8
Warehousing	604	132	21.9

NOTE: Counts include all sampled and responded companies except for ineligibles. Response counts include blank records for subsidiaries that were included and named by their parent company and for companies that contacted us to tell us that they had no computers, had no Internet access, or outsourced all IT.

Data-Editing Protocols

With respect to data editing, both range and edit checks were incorporated into the Web form and optical scanning (OPSCAN) form. Edit checks and logical imputations also were conducted. Deduplication of survey responses was also done for companies that submitted multiple surveys. We also summarize how the survey data were handled for companies that filled out the survey and included information about their subsidiary firms within their corporate structure. Appendix A presents a detailed summary of the data-editing protocols and checks done for the NCSS.

CHAPTER FIVE

Weighting Methodology and Nonresponse Analysis

Missing Data

In every survey, two types of missing data can arise: unit nonresponse and item nonresponse. *Unit nonresponse* occurs when the sampled unit either is not found or declines to participate. *Item nonresponse* occurs when a sampled unit responds to only some of the survey questions. It is customary to deal with the missing data arising from unit nonresponse with weights and with the missing data that arise from item nonresponse via imputation.

In agreement with the sponsor, we adopted this strategy, described further in Appendix A, limiting the imputation exercise only to logical imputation (as described next). Given the low unit-response rate and the nature of most of the questions in the survey instrument (i.e., questions that ask the respondent to "mark all that apply" of a list of possible answers), we decided not to perform imputations for certain types of item nonresponse. Similar to what was done for the pilot survey, we decided to perform logical imputation. For example, all the subsections in Section III of the survey instrument (types of computer-security incidents) start with a screener question asking whether that type of incident has been detected. If that item was missing but the respondent supplied a positive response to the financial-loss questions elsewhere in that subsection, then the screener question was imputed with a "yes." Other similar logical imputations were implemented. See Appendix A for a detailed summary of the logical imputations done.

Nonresponse Analysis

Before describing in detail how we dealt with unit nonresponse, we briefly discuss the different typologies of missing data and their consequences on estimates. Rubin (1976) developed a typology or terminology for missing data, or, more precisely, for the distribution of missingness. Such typology or terminology is widely used even though somewhat confusing and widely misunderstood. Rubin identified three different types of missing data: missing completely at random (MCAR), missing at random (MAR), and missing not at random (MNAR). We describe these three types in very general terms.

MCAR means that the probability that a certain outcome is missing for a given subject does not depend either on the observed characteristics of the subject or on the outcome itself for this subject. If the MCAR assumption is correct, then the missing data do not bias estimates, but just have a consequence on the precision of the estimates due to the loss of sample size.

MAR means that the probability that a certain outcome is missing for a given subject depends on some or all the observed characteristics of the subject but not on the outcome itself for this subject. If the MAR assumption holds and nothing is done to account for it, the estimates obtained using only the observed data are likely to be biased in addition to being less accurate. In the case of the NCSS, it is likely that larger companies have a lower response rate than smaller companies. It is also likely that the response rate varies by industry. Both the industry and company-size variables are observed or known both for the responders and for the nonresponders.

MNAR means that the probability that a certain outcome is missing for a given subject depends on the outcome itself. If the MNAR assumption holds, then the estimates obtained using the observed data will be biased (in addition to being less precise). In the case of the NCSS, it might be that the companies that have experienced a lot of cyberattacks tend to respond at a lower rate. However, if these differential response rates are explained by industry or company size, because only certain industries and companies of a certain size get attacked, the MNAR assumption will not hold after all and the MAR assumption is what we need.

MCAR, MAR, and MNAR are assumptions, and, as Schafer and Graham (2002) stressed, "there is no way to test whether MAR, for example, holds in a data set, except by obtaining follow-up data from non-respondents." Schafer and Graham (2002) also argued that the MAR assumption is quite plausible in many situations. Even when the missing data have a distribution that does not satisfy the MAR assumption entirely, "Collins, Schafer and Kam (2001) demonstrate that in many realistic cases, an erroneous assumption of MAR may often have only a minor impact on estimates and standard errors."

In what follows, we make the assumption that the missing data arising from unit nonresponse in NCSS follow the MAR assumption. This is the only feasible assumption to make for the NCSS data, given that all the techniques that deal with data that are assumed to be MNAR require the specification of a distribution for the missingness and of a model for the complete data. In other words, these methods require knowing beforehand what kind of analyses any potential analyst might want to do with the NCSS data. This, of course, is impossible, and we leave the freedom to make such an assumption to the potential users of the NCSS data.

However, we provide the general user of the NCSS data with nonresponse weights, built as a function of the interaction of the industry and company-size variables. In short, the weights derive from the probabilities of response, which we estimate from the data. Weighting can remove bias that is due to differential response rates associated with the variables used to estimate the response probabilities and nonresponse weights. Therefore, nonresponse weights work under the MAR assumption. Note that weights do not correct for biases related to unused or unobserved variables. In the NCSS case, the only variables that we observe for both the responders and nonresponders are industry and company size. Weighting has several advantages. In particular, it is nonparametric; therefore, it does not require assuming any particular model for the complete data. In other words, it is a very versatile way to handle missing data. In addition, weights are easy to use, since modern statistical software packages allow the inclusion of weights.

Table 4.4 in Chapter Four presented a detailed summary of response rates for each industry sector, by risk category, employment size, and certainty companies. The nonresponse analyses were limited by the lack of information on the nonrespondents, which was restricted to the basic sampling frame data. We choose to look at nonresponse across industry, across employee size, and as an interaction of these two variables. Weighted tabulations of these showed highly significant differences in response across industry groups ($\chi2$ = 86.55, p = <0.0001) and across size groups ($\chi2$ = 312.49, p = <0.0001) when looked at individually. Industries with the lowest and highest response rates are presented in Tables 5.1 and 5.2, respectively.

The smallest companies (2–9 employees) had the lowest response rate at 15.4 percent, whereas the other size groups' response rates ranged from 20.4 percent to 32.8 percent. Crossing industry and size to look at the interaction of the two variables revealed that company size affected response in some industries and, in others, had no effect. The tabulation of response by industry and size showed significant differences in response across the strata or industry and size ($\chi2$ = 754.3, p = <0.0001). These results stress the importance of using the nonresponse weights when conducting analyses of the NCSS data to reduce the amount of bias as much as possible.

Table 5.1
Industries with the Lowest Response Rates

Industry	Response Rate (%)
Picture and Sound Recording	13.9
Accommodations	14.2
Telecommunications	16.6
Real Estate	18.4
Food Services	18.9

Table 5.2
Industries with the Highest Response Rates

Industry	Response Rate (%)
Utilities	38.1
Manufacturing, Durable	27.9
Accounting and Bookkeeping	26.2
Forestry, Fishing & Hunting	24.2
Architecture, Engineering, and Specialized Design Services	24.2

Weighting Scheme

Since we adopted a stratified sampling design and the number of sampled companies in every stratum was not selected proportional to the stratum size, companies in different strata were sampled with different probabilities. The reason for doing so was being able to produce accurate estimates not only at the national level, but also at the industry and company-size levels. This implied having to oversample certain strata versus other strata. Even if we had sampled all the companies with the same probability, companies are likely to be included in the sample with differential probabilities because of nonresponse.

When subjects are sampled with differential probabilities, if nothing is done to account for these differential probabilities, the estimates will be biased. The way in which such bias is eliminated is by weighting, where the weights are the inverse of the inclusion probabilities. Several types of weights have been generated for the NCSS: sampling weights, nonresponse weights, analytic or final weights, and post-stratification weights.

In what follows, we describe each of the different types of weights.

Sampling Weights

As mentioned, because of the adopted sampling design and allocation, companies were sampled with differential probabilities. Generally, companies belonging to the same stratum were sampled with the same probability, while companies belonging to different strata were sampled with different sampling probabilities. The exception to this rule is those strata that contain both certainty and noncertainty companies. To account for the differential sampling probabilities, we built sampling weights, which are given by the inverse of the sampling probabilities.

Nonresponse Weights

The final probability of a company's being in the sample is given by the product of the sampling probability (or the probability of being selected) times the probability of responding, given that the company was selected to be part of the sample. We computed the response probability as a function of industry and company size. In other words, the response probability is given by the ratio of the number of companies that responded in a given stratum divided by the number of companies allocated to that stratum. This computation is equivalent to fitting a logistic regression where the outcome variable is the response stratum for every company in the sample, predicted by the interaction of the industry and company-size variable. Again, the nonresponse weights are given by the inverse of the response probabilities.

Analytic Weights

The analytic weight is given by the inverse of the final probability of a company to be included in the sample. This probability is the product of the sampling/selection probability times the probability of responding, given that the company was sampled. The analytic weight is obtained either by taking the product of the sampling weight and nonresponse weight or by taking the ratio of the total number (in the population) of companies in a stratum divided by the number of companies that responded in that stratum. The analytic weights were normalized so that they summed to the total number of companies in the frame.

Post-Stratification Weights

In addition to the sampling and nonresponse weights, we generated post-stratification weights.

As discussed previously, among the strata of our sampling design, we had created nine certainty strata: one for every company-size class. In other words, the certainty companies of a given size, independently of which industry to which they belong, were grouped together in the certainty stratum of that company-size class. This design decision was sensible, since it was thought that certainty companies had the same importance no matter which industry they belonged to, and, thus, it made sense to place the certainty companies in their own stratum. Also, this design decision was made under the assumption of attaining a 50 percent or higher response rate. Given the response pattern of the certainty companies, when computing the weighted frequencies by industry, these frequencies were not found to match the population counts. Therefore, we generated post-stratification weights to solve this problem. That is, post-stratification weights were needed to allow generation of industry-level counts that would match the industry counts in the frame.

From the strata of the original design, we eliminated the certainty strata, and we also collapsed the last three employment-size classes. This collapsing was necessary, because, for some of the large company-size classes and certain industries, no certainty company responded, leaving some strata (where these strata are defined by, for example, industry x and company-size class 9) with no respondents. This happened because certain strata, like the one described earlier, contained only certainty companies.

Therefore, the post-strata are defined by the interaction of the industry variable, with 36 levels, and a new employment-size variable with seven levels. This means that we are left with 252 strata or post-strata.

To compute the post-stratification weights, we first assigned the certainty companies to their own industry and size strata. Then we computed the percentage of companies in the population and the weighted (using the analytic weights) percentage of the companies in the sample falling into the post-strata. The post-stratification weights are given by the ratio of the first percentage divided by the second one.

When conducting analyses, the weight to be used is given by the product of the post-stratification weights times the analytic weights defined previously.

Examples of Weighted Estimates and Standard Errors

Presented here are three parts to an example on generating weighted estimates and standard errors using the SUDAAN software package. The parts include SUDAAN code, SUDAAN output, and a table presenting the data generated by the code and output.

The following SUDAAN code generates basic univariate statistics, such as estimates, sample sizes, and standard errors, for nine variables from the survey. The important details in the code are the design-specification options, which include the sample design method (design=strwor), the sampling stratification variable (nest stratnum), weighting variable (post_awt), and the total population counts (totcnt). These statements will be used with any of the SUDAAN procedures when analyzing the NCSS data.

```
proc descript data=ncss filetype=SAS design=strwor;
    nest stratnum / missunit;
    weight post_awt;
    totcnt fnco;
    var q5ax q5bx q6ax q7ax q8ax q9ax q10ax q11ax q12ax q13x;
    catlevel 1 0 1 0 1 0 1 0 1 0 1 0 1 0 1 0 1 0 1 0;
    print nsum wsum total setotal percent sepercent lowpct uppct;
run;
```

The design method, *Stratified Without Replacement (design=strwor)*, is for single-stage sampling designs that use stratified random sampling without replacement and have equal probabilities of selection within each stratum. The NEST statement is required, and, in the case of the strwor design method, identifies the stratification variables used in the sampling. In the code shown, a single variable has been created to indicate the strata; however, one could also use the actual sampling stratification variables indgrpb (industries including certainty companies as a stratum) and empcat (employee-size categories). The MISSUNIT option tells SUDAAN how to calculate standard errors in the event that there is one or fewer observation in any stratum. The WEIGHT statement identifies the variable that is the analysis weight and is a required statement as well. The TOTCNT statement is also required with this design and is simply the total population counts for each sampling stratum. The counts must correspond to the variable(s) used in the nest statement, e.g., if you specify nest stratnum, a single variable for the population counts can be used in the totcnt statement. If nest indgrpb empcat is used, two variables for the population counts must be used. The remaining statements are specific to the procedure and can be found in the SUDAAN language manual.

The SUDAAN output generated by the sample code is as follows, and Table 5.3 has presented results:

EXAMPLE OF WEIGHTED ESTIMATES AND STANDARD ERRORS USING SUDAAN

```
                            S U D A A N
            Software for the Statistical Analysis of Correlated Data
         Copyright       Research Triangle Institute        July 2004
                            Release 9.0.0


Number of observations read     :    7889    Weighted count :   7278109
Number of observations skipped :     300
(WEIGHT variable nonpositive)
Denominator degrees of freedom :    7938

        Date: 12-19-2007                          Research Triangle Institute
        Page : 1
        Time: 15:58:03                            The DESCRIPT Procedure
        Table : 1

Variance Estimation Method: Taylor Series (STRWOR)
by: Variable, One.
```

| Variable | | One |
| | | 1 |

Q5A-W/LOGICAL	Sample Size	7367
IMPUTATION: 1	Weighted Size	5518444.10
	Total	2932581.43
	SE Total	156390.99
	Percent	53.14
	SE Percent	2.47
	Lower 95% Limit	
	Percent	48.28
	Upper 95% Limit	
	Percent	57.95

Q5B-W/LOGICAL	Sample Size	7311
IMPUTATION: 1	Weighted Size	5217495.48
	Total	1541599.71
	SE Total	122120.99
	Percent	29.55
	SE Percent	2.19
	Lower 95% Limit	
	Percent	25.45
	Upper 95% Limit	
	Percent	34.00

Q6A-W/LOGICAL	Sample Size	7297
IMPUTATION: 1	Weighted Size	5350835.66
	Total	684869.32
	SE Total	83780.11
	Percent	12.80
	SE Percent	1.53
	Lower 95% Limit	
	Percent	10.08
	Upper 95% Limit	
	Percent	16.12

Q7A-W/LOGICAL	Sample Size	7277
IMPUTATION: 1	Weighted Size	5412341.70
	Total	105106.12
	SE Total	35795.99
	Percent	1.94
	SE Percent	0.66
	Lower 95% Limit	
	Percent	0.99
	Upper 95% Limit	
	Percent	3.76

```
        Date: 12-19-2007                        Research Triangle Institute
        Page  : 2
        Time: 15:58:03                          The DESCRIPT Procedure
        Table : 1
```

Variance Estimation Method: Taylor Series (STRWOR)
by: Variable, One.

```
------------------------------------------------------
| Q8A-W/LOGICAL  | Sample Size       |        7272 |
| IMPUTATION: 1  | Weighted Size     |  5365319.80 |
|                | Total             |    80550.32 |
|                | SE Total          |    35141.05 |
|                | Percent           |        1.50 |
|                | SE Percent        |        0.65 |
|                | Lower 95% Limit   |             |
|                |  Percent          |        0.64 |
|                | Upper 95% Limit   |             |
|                |  Percent          |        3.50 |
------------------------------------------------------
|                |                   |             |
| Q9A-W/LOGICAL  | Sample Size       |        7269 |
| IMPUTATION: 1  | Weighted Size     |  5388112.08 |
|                | Total             |   156296.98 |
|                | SE Total          |    41007.18 |
|                | Percent           |        2.90 |
|                | SE Percent        |        0.75 |
|                | Lower 95% Limit   |             |
|                |  Percent          |        1.74 |
|                | Upper 95% Limit   |             |
|                |  Percent          |        4.81 |
------------------------------------------------------
|                |                   |             |
| Q10A-W/LOGICAL | Sample Size       |        7275 |
| IMPUTATION: 1  | Weighted Size     |  5341022.04 |
|                | Total             |    40122.06 |
|                | SE Total          |     7788.33 |
|                | Percent           |        0.75 |
|                | SE Percent        |        0.15 |
|                | Lower 95% Limit   |             |
|                |  Percent          |        0.51 |
|                | Upper 95% Limit   |             |
|                |  Percent          |        1.10 |
------------------------------------------------------
|                |                   |             |
| Q11A-W/LOGICAL | Sample Size       |        7258 |
| IMPUTATION: 1  | Weighted Size     |  5424084.98 |
|                | Total             |    39294.36 |
|                | SE Total          |    11514.46 |
|                | Percent           |        0.72 |
|                | SE Percent        |        0.21 |
|                | Lower 95% Limit   |             |
|                |  Percent          |        0.41 |
|                | Upper 95% Limit   |             |
|                |  Percent          |        1.29 |
------------------------------------------------------
```

```
        Date: 12-19-2007                        Research Triangle Institute
        Page  : 3
        Time: 15:58:03                          The DESCRIPT Procedure
        Table : 1
```

Variance Estimation Method: Taylor Series (STRWOR)
by: Variable, One.

```
------------------------------------------------------
|                |                   |             |
| Variable       |                   | One         |
|                |                   | 1           |
------------------------------------------------------
|                |                   |             |
| Q12A-W/LOGICAL | Sample Size       |        7261 |
| IMPUTATION: 1  | Weighted Size     |  5350047.99 |
|                | Total             |   719218.12 |
|                | SE Total          |    95396.33 |
|                | Percent           |       13.44 |
|                | SE Percent        |        1.74 |
|                | Lower 95% Limit   |             |
|                |  Percent          |       10.39 |
|                | Upper 95% Limit   |             |
|                |  Percent          |       17.23 |
------------------------------------------------------
```

Table 5.3
Weighted Estimates and Standard Errors Presenting the Results of the SUDAAN Output

Computer-Security Incident Type	Weighted Percent	Standard Error of Percent	Survey Sample Size	Estimated Population Count	Standard Error of Estimate Population Count
Computer Virus Intercepted Prior to Infection	53.1	2.5	7,367	2,932,581	156,391
Computer Virus Detected After Infection	29.5	2.2	7,311	1,541,600	122,121
Denial of Service	12.8	1.5	7,297	684,869	83,780
Electronic Vandalism or Sabotage	1.9	0.7	7,277	105,106	35,796
Embezzlement	1.5	0.7	7,272	80,550	35,141
Fraud	2.9	0.7	7,269	156,297	41,007
Theft of Intellectual Property	0.7	0.2	7,275	40,122	7,788
Theft of Personal or Financial Information	0.7	0.2	7,258	39,294	11,514
Other Computer Security Incident	13.4	1.7	7,261	719,218	95,396

Range and Edit Checks and Other Data Cleaning Procedures

This appendix is organized in the following sections:

A. Range and edit checks incorporated into the web form and optical scanning (OPSCAN) form.

B. Edit checks and logical imputations conducted for data cleaning and editing.

C. Survey responses that included subsidiaries.

D. De-duplication of survey responses for companies that submitted multiple surveys.

RANGE AND EDIT CHECKS INCORPORATED INTO THE WEB FORM AND OPSCAN.

1) Check Box Questions with a Limit.

The questions in this category are: 1a, 1b

For the OPSCAN, we recorded as data *all* boxes that were checked, even if they exceed the stated limit. For the Web version, a maximum of three checks was imposed. All fill-in-the-blank information was retained, even if the corresponding check box was not checked.

The Web form has been programmed so that there is no limit to how much "written" information a respondent can enter. On the paper/scannable form, if there was a written response in the "Other→Specify" field—the person responsible for verification would type in the response. Therefore, all of the written responses in these fields would be captured.

2) Check Box Questions without a Limit.

The questions in this category are: 2a, 2b, 3a, 3d, 4a, 4b, 4c, 4d, 5c, 5d, 5e, 5g, 5h, 6b, 6c, 6d, 6f, 6g, 7b, 7c, 7d, 7f, 7g, 8b, 8c, 8d, 8f, 8g, 9b, 9c, 9d, 9f, 9g, 10b, 10c, 10d, 10e, 10g, 10h, 11b, 11c, 11d, 11e, 11g, 11h, 12b, 12c, 12d, 12e, 12g, 12h, 14, 15, 17, 21

- Record as data *all* boxes that were checked.
- Record as data any fill-in-the-blank information, even if the corresponding check box isn't checked.

- Accept any character (except blank/space) as a valid check mark; it doesn't have to be "x" or "X".
- If a "None" or "Not applicable" or "Never" or "Don't know" box is checked *in addition to* other checkboxes,
  - If the form is being filled in interactively, query the user about the discrepancy
  - If not interactive (paper), ignore the "None, etc." checkbox if there is other valid information checked or written in

The Web form has been programmed so that there is no limit to how much "written" information a respondent can enter.

3) Yes/No Type Questions.

Questions in this category are: 4e, 5a, 13, and 15

Edit checks programmed into the web form for these questions were:

- If one box is checked, record that information.
- If neither box is checked, but the (optional) numeric field is filled in, treat as a "yes" and record that numeric data;
- If two or more of "yes" and "no" and (for certain questions) "don't know" are checked and the form is being filled in interactively, query the user about the discrepancy.
- For question 13, allow up to six digits for an integer response to "How many incidents..."

4) Yes/No Screener Questions.

These questions are at the beginning of subsections, to determine if the questions in the subsection are applicable.

Questions in this category are: 5b, 6a, 7a, 8a, 9a, 10a, 11a, and 12a.

Edit checks programmed into the Web Form for these questions were:

- If a "no" response is given to a screener question, the subsequent questions in that sub-section are skipped.
- For the OPSCAN, this could not be automated, thus it was made part of the data cleaning specifications.

5) Percentage Questions.

Questions in this category are: 3c, 16, 18b

For these questions, a valid non-blank answer is an integer from 0 to 100 inclusive. Allow three digits for an integer response. Blank is treated as "no response".

For the OPSCAN equipment, any additional marks such as a "period" will come up as

"unrecognizable" by the OPSCAN equipment. These will be reviewed by the verifier and corrected to ensure accurate capture of this data.

6) "Number of Incidents" Questions. (number of incidents were detected, number that were reported, etc.)

Questions in this category are: 5f, 6e, 7e, 8e, 9e, 10f, 10k, 11f, and 12f.

Edit checks programmed into the Web Form for these questions were:
- For these questions, the response allows for six digits
- Only written zeroes should be treated as zero. Blanks should be treated as missing.

In addition:
- the answer in 5f should be less than or equal to the answer in 5b
- the answer in 6e should be less than or equal to the answer in 6a
- the answer in 7e should be less than or equal to the answer in 7a
- the answer in 8e should be less than or equal to the answer in 8a
- the answer in 9e should be less than or equal to the answer in 9a
- the answers in 10f and 10k should each be less than or equal to the answers in 10a
- the answer in 11f should be less than or equal to the answer in 11a
- the answer in 12f should be less than or equal to the answer in 12a

If any of the above conditions are not true, the response is flagged.

7) Number of Hours Questions.

Questions in this category are: 5i, 6h, 7h, and 12i.

- The answer should be a non-negative integer.
- Blank is treated as a non-response.
- Six digits are allowed for the response.

8) Dollar Value Questions.

These questions ask for a dollar value, separated into millions, thousands, and hundreds (hard-coded on the form). Questions in this category are: 3b, 5j, 5k, 6i, 6j, 7i, 7j, 8h, 8i, 9h, 9i, 10i, 10j, 11i, 11j, 12j, 12k, 18a

- Divide all dollar value variables by 1000 to make consistent with the paper data.[11]
- All of the "thousands" data should be an integer between 0 and 999.
- In question 18a, since there's a billions field, the "millions" integer should be between 0 and 999.
- For the billions field, allow two digits.
- When there is no billions field, the millions field can be larger than 999; in that case, allow 5 digits for the millions field.

For the 16 dollar values listed in the table below (5j thru 12k) all values are allowed subject to the number of digits specified above.

Incident type	Dollars spent to recover	Other monetary losses and costs
5 virus	5j	5k
6 denial of service	6i	6j
7 vandal/sabotage	7i	7j
8 embezzlement	8h (value taken)	8i
9 fraud	9h (value taken)	9i
10 theft of IP	10i (value taken)	10j
11 theft of personal info	11i (value taken)	11j
12 other	12j	12k

9) Other Data Questions.

Question 19 (number of employees).

- Allow 9 digits for the response.
- Blank is treated as a non-response.

Question 20 has check boxes for calendar year 2005 or fiscal year 2005 (only one should be checked). If box 20.02 is checked the month(s) should be an integer

---

[11] Due to a programming error, the dollar values had to be divided by 1000 to correct for the web vendor mistakenly multiplying these values by 1000.

from 1 to 12, and the year should be either a 4-digit number from 2003 to 2006, or else a two-digit number from 03 to 06 (which should be interpreted as 2003 to 2006, respectively). The second date shouldn't exceed the current date. If the second date is less than the first date, the response is flagged for human review.

Question 21 has two check boxes. Exactly one should be checked. If the second box is checked, a 2-digit integer should appear in the "number of subsidiaries" box.

EDIT CHECKS AND LOGICAL IMPUTATIONS CONDUCTED FOR DATA CLEANING AND EDITING

General Edits:

1) All dollar value variables are expressed in thousands (1000s). Variables affected are: q3b, q5j, q5k, q6i, q6j, q7i, q7j, q8h q8i, q9 h, q9i, q10i, q10j, q11i, q11j, q12j, q12k, q18a.

2) All percentage variables have range checks. Values below 0 are set to 0; values above 100 are set to 100. Variables affected are: q3c, q16, q18b.

3) Inconsistencies between Section III screener questions and the subsequent non-trivial data are resolved. Data reported for at least one of the following concepts: count of incidents, dollars spent to recover, or other monetary losses or costs is considered a non-trivial response. (Example: if q5b not equal 1 and data is provided in the follow-up questions q5bcnt or q5j or q5k, then we logically impute q5b=1). Screener variables affected are: q5b, q6a, q7a, q8a, q9a, q10a, q11a, q12a.

4) Logically checked that the number of incidents reported is less than or equal to the number of incidents detected. (Example: q5f <= q5bcnt). If the number reported is greater than the number detected, then the number reported is capped at the value of the number detected. Variables affected: q5f, q6e, q7e, q8e, q9e, q10f, q11f, and q12f.

5) If they do not report how many incidents were detected, however they do report how many incidents were reported to agencies, then the number detected is imputed to be equal to the number reported. Variables for number detected are: q5bcnt, q6acnt, q7acnt, q8acnt, q9acnt, q10acnt, q11acnt, q12acnt.

6) If the only response for which organization they reported the detected incidents to is "None of the Above" (q5e9=1), then the number of incidents reported is logically imputed to zero. Variables for number reported are: q5f, q6e, q7e, q8e, q9e, q10f, q11f, q12f.

7) If any open-ended response question is completed, but its respective check box is not checked we logically impute the check box =1.

8) If a "check-all-that-apply"" question has a response of none, not applicable, or don't know, as well as other boxes checked, then logically impute this check box to zero. Variables affected are: q3d, q4b, q4c, q4d, q5c, q5d, q5e, q6b, q6c, q6d, q7b, q7c, q7d, q8b, q8c, q8d, q9b, q9c, q9d, q10c, q10d, q10e, q11c, q11d, q11e, q12c, q12d, and q12e.

Variable Specific Edits:

1) Question 4d: If the response to q4d indicates that a vulnerability/risk assessment was conducted, but none are listed in questions q4a or q4b, then we impute q4a to be consistent with q4d.

2) Question 4a: Anything reported in question q4b should also be reported in q4a. Check boxes in q4a are imputed to reflect boxes checked in q4b.

3) Question 10k: A logic check was made to insure that the number of incidents detected involving digital products is less than or equal to the number of total incidents detected (q10k <= q10acnt). If the logic check failed (q10k > q10acnt) then digital products incidents are capped at the total number of incidents of theft of intellectual property (if q10k > q10acnt then q10k=q10acnt).

4) Question 13: If there was inconsistency between the dichotomous y/n variable about stolen laptop incidents and number of incidents reported, then the yes/no variable was imputed appropriately   Example: If q13 is not equal to 1 and q13cnt>0 then logically impute q13=1.

5) Question 20: If q20 is blank and dates are provided, then q20 is logically imputed to be equal to "Fiscal year or other time period" (q20=2).

6) Question 20: Dates reported are edited to a consistent format. If they indicated a reporting period other than calendar year then all dates reported are logically imputed to a consistent format.

7) Question 21: If q21 is blank and the number of subsidiaries is non-zero (q21a>0), then q21 is imputed to be yes. Example: if q21 is null and q21a >=1 then q21=1.

SURVEY RESPONSES THAT INCLUDED SUBSIDIARIES

In some cases, companies filled out the survey including information for subsidiary firms within their corporate structure. We determined this from the response to question q21: "Information includes this company and some subsidiaries." We obtained a list of these subsidiaries and compared these to the survey frame to determine if any of the subsidiaries were part of the NCSS sample. If so, we included records for those subsidiaries in the database. We did not, however, fill in any of the fields for these subsidiary records. This is because we could not determine how to distribute the responses of the mother company down to the level of their subsidiaries. Therefore, all the fields on the subsidiary records are blank, except for the sampling strata variables.

To identify these records in the analytic dataset there are 2 variables added to the file, sublink and subflag. Sublink links the subsidiary records to the responding company (donor) that indicated their report also included data on those subsidiary companies (donee). The donor and donee records all have the same value for sublink. For example, sublink=105 links itself with three subsidiary records included in their report. The variable subflag can be used to differentiate the donor and donee records. Subflag=0 indicates the record represents the response donor and subflag=1 indicates the donee record, which was added to the dataset. If subflag is missing then there are no subsidiary records linked to it in the dataset.

Inclusion of sampled subsidiaries in the dataset was done by four methods each with decreasing levels of reliability. The majority (68%) of subsidiaries were included by matching of company names. This was done by matching the names of subsidiaries that the parent companies indicated they included in their survey report to the sampled companies for that parent.

For companies that indicated their report did include subsidiaries, however failed to list those subsidiaries by name, we made count comparisons. The count comparisons were made by using three criteria. The count of total number of subsidiaries belonging to a parent company was calculated using the D&B sampling frame data. If the D&B count was equal to the total number of subsidiaries a parent company indicated including in the survey responses, then the sampled subsidiaries of this parent were included in the database. This count match accounted for 6% of the added subsidiaries. If the parent company's total number of subsidiaries included in the survey report was greater than the D&B total count of subsidiaries for that parent the sampled subsidiaries were included in the database. Counts that differed by 20% or less (5.7%) were distinguished from counts that differed by greater than 20% (20.3%). The method used to select sampled subsidiaries for inclusion in the database is indicated in the variable SUBMATCH (1 is name match, 2 is N reported=N D&B, 3 is N reported <=20% greater than D&B Count, 4 is N reported >20% greater than D&B Count).

Analysts will need to make their own decisions if and how to distribute the information on the parent company record to the subsidiary records. Consider the following simple example: A parent company has three blank subsidiary records representing the companies in the sample frame for which the parent company provided data. If one assumes that the information was evenly distributed across the four companies (the parent company plus the three subsidiaries), then the counts of incidents detected, incidents reported, the dollar costs of the incidents, etc, could be apportioned out such that each company received 25% of the total. This is one very simple example of how the information supplied on behalf of subsidiary companies could be allocated to those companies.

DE-DUPLICATION OF SURVEY RESPONSES FOR COMPANIES THAT SUBMITTED MULTIPLE SURVEYS

In some cases, multiple responses were received from the same company. In this situation, the multiple responses were consolidated into a single response. The first response received was treated as the "main response row" and it was supplemented with additional information from later responses to create a single record for each firm that is as complete as possible. First response was determined based on the time stamp on the completed web survey, or the date when a paper form was received in the mail. One reason for choosing the "first" versus "last" response is that we know in some cases, respondents eventually figure out that if they answer "no" to all of the screener questions in Section III they can get to the end faster.

APPENDIX B
Questionnaire

FORM **NCSS-1**

(4/24/06)

OMB No. 1121-0301: Approval Expires 08/31/2008

2005 NATIONAL COMPUTER SECURITY SURVEY

RETURN COMPLETED FORM TO:
RAND Corporation
Survey Research Group
1776 Main Street
P.O. Box 2138
Santa Monica, CA 90407-2138

OR
FAX TO:
1-877-814-6673

For assistance
Phone: 1-800-734-5399
Monday through Friday
8:00 a.m. to 5:00 p.m. Pacific Time
OR
E-mail: ncss@rand.org

 U.S. DEPARTMENT OF JUSTICE
BUREAU OF JUSTICE STATISTICS

In partnership with the

 U.S. DEPARTMENT OF HOMELAND SECURITY
NATIONAL CYBER SECURITY DIVISION

SURVEY SCOPE
This voluntary survey collects data on the type and frequency of computer security incidents in which a computer was used as the means of committing a crime against the company.

REPORTING ENTITY
Report consolidated figures for DOMESTIC OPERATIONS of this company, including all DIVISIONS and LOCATIONS, and **excluding** SUBSIDIARIES. *Use figures that include subsidiaries only if figures excluding subsidiaries are not available.* For this survey, subsidiary means a company in which this company has more than 50% ownership, or in which this company has the power to direct or cause the direction of management and policies.

REPORTING PERIOD
The reporting period for this survey is CALENDAR YEAR 2005. If 2005 calendar year figures are not available, please use fiscal year 2005 data.

ESTIMATES
If exact figures are not available, estimates are acceptable.

Use a dark colored pen to fill out the survey. Completely fill in the squares ■ or circles ● to indicate your responses. To indicate an answer selected in error, draw a heavy "X" over the square or circle. When reporting a number, avoid writing on the edge of the response box. **Please refer to the instructions on page 14 before completing the survey.**

I. COMPUTER SECURITY CONCERNS

1a. What are the top three computer security concerns for this company? *Mark ■ up to three.*

☐ Computer virus, worm, or Trojan horse

☐ Denial of service

☐ Electronic vandalism or sabotage

☐ Embezzlement

☐ Fraud

☐ Theft of intellectual property (copyrights, patents, trade secrets, trademarks)

☐ Unlicensed use or copying (piracy) of digital products—software, music, motion pictures, etc.—developed for resale

☐ Theft of personal or financial information such as names and dates of birth; social security numbers; credit/debit/ATM card, account, or PIN numbers; etc.

☐ Other computer security incidents such as hacking, spoofing, phishing, sniffing, pinging, scanning, spyware, adware, other malware, etc.

☐ Misuse of computers by employees (Internet, e-mail, etc.)

☐ Breaches resulting from information obtained from stolen laptops

☐ Other⟶ Specify: _____

b. What three potential sources of computer security threat are of greatest concern to this company? *Mark ■ up to three.*

☐ Current employee

☐ Current contractor, vendor, temporary worker, etc.

☐ Former employee, contractor, vendor, temporary worker, etc.

☐ Domestic competitor

☐ Foreign competitor

☐ Domestic hacker

☐ Foreign hacker

☐ Other⟶ Specify: _____

II. COMPUTER INFRASTRUCTURE & SECURITY

2a. In 2005, what types of computer networks (including Internet) or equipment did this company use?
For this survey, "company" means DOMESTIC OPERATIONS, including all DIVISIONS and LOCATIONS. *Mark ■ all that apply.*

☐ Local area network (LAN)

☐ Wide area network (WAN)

☐ Process control network (PCN)

☐ Virtual private network (VPN)

☐ Wireless network (e.g.,802.11)

☐ Electronic data interchange (EDI)

☐ Internet

☐ Intranet

☐ Extranet

☐ Stand-alone PCs (not on LAN)

☐ Company-owned laptops

☐ Laptops not owned by company

☐ Other⟶ Specify:

_____

b. In 2005, what types of network access did this company support? *Mark ■ all that apply.*

☐ Hard-wired telecommunications lines

☐ Remote dial-in access via telecommunications lines

☐ Access to company networks or e-mail through Internet

☐ Wireless access to e-mail

☐ Wireless access to Internet

☐ Wireless access to this company's data or other networks

☐ Publicly accessible website WITHOUT e-commerce capabilities

☐ Publicly accessible website WITH e-commerce capabilities

☐ Other⟶ Specify: _____

13532

II. COMPUTER INFRASTRUCTURE & SECURITY - Continued

3a. In 2005, what types of computer system security technology did this company use? Mark ■ all that apply.

- ☐ Anti-virus software
- ☐ Anti-spyware/adware software
- ☐ Biometrics
- ☐ One-time password generators (smartcards, tokens, keys)
- ☐ Passwords that must be changed periodically
- ☐ Digital certificates
- ☐ Firewall
- ☐ DMZ Host
- ☐ Intrusion Detection System
- ☐ Intrusion Protection System
- ☐ E-mail logs or filters
- ☐ System administrative logs
- ☐ Encryption
- ☐ Other ⟶ Specify: _____

b. In 2005, how much did this company spend on the types of computer system security technology identified in 3a? ESTIMATES are acceptable. EXCLUDE personnel costs.

	Mil.		Thou.		Dol.
$					0 0 0

c. What percentage of this company's total 2005 Information Technology budget did this company spend on the types of computer system security technology identified in 3a? ESTIMATES are acceptable. *Round to nearest whole percent.*

[___] %

d. What types of computer system security technology does this company plan to add in 2006? EXCLUDE updates or upgrades of technologies already used in 2005. Mark ■ all that apply.

- ☐ Anti-virus software
- ☐ Anti-spyware/adware software
- ☐ Biometrics
- ☐ One-time password generators (smartcards, tokens, keys)
- ☐ Passwords that must be changed periodically
- ☐ Digital certificates
- ☐ Firewall
- ☐ DMZ Host
- ☐ Intrusion Detection System
- ☐ Intrusion Protection System
- ☐ E-mail logs or filters
- ☐ System administrative logs
- ☐ Encryption
- ☐ Other ⟶ Specify: _____
- ☐ Do not plan to add any new technologies in 2006

4a. In 2005, what types of computer security practices did this company have? Mark ■ all that apply.

- ☐ Business continuity plan for computer systems
- ☐ Disaster recovery plan for computer systems
- ☐ Corporate policy on computer security
- ☐ Identification of company's critical assets
- ☐ Vulnerability/risk assessment
- ☐ Intrusion/penetration testing of computer security
- ☐ Computer/network watch center
- ☐ Configuration management
- ☐ Regular review of system/security administration logs
- ☐ Periodic computer security audits
- ☐ Formal computer security audit standards
- ☐ Physical/environmental security (e.g., limited physical access, sprinklers)
- ☐ Personnel policies (e.g., background checks, transfer, termination)
- ☐ Training employees in computer security practices
- ☐ Equipment decommissioning
- ☐ Other ⟶ Specify: _____

b. In 2005, what computer security functions did this company outsource? INCLUDE fully and/or partially outsourced functions. Mark ■ all that apply.

- ☐ Business continuity plan for computer systems
- ☐ Disaster recovery plan for computer systems
- ☐ Corporate policy on computer security
- ☐ Identification of company's critical assets
- ☐ Vulnerability/risk assessment
- ☐ Intrusion/penetration testing of computer security
- ☐ Computer/network watch center
- ☐ Configuration management
- ☐ Regular review of system/security administration logs
- ☐ Periodic computer security audits
- ☐ Formal computer security audit standards
- ☐ Physical/environmental security (e.g., limited physical access, sprinklers)
- ☐ Personnel policies (e.g., background checks, transfer, termination)
- ☐ Training employees in computer security practices
- ☐ Equipment decommissioning
- ☐ Other ⟶ Specify: _____
- ☐ None; all computer security was done in-house

c. If this company had a computer system business continuity or disaster recovery plan, was it tested, used in an emergency situation and/or updated in 2005? Mark ■ all that apply.

- ☐ Tested in 2005
- ☐ Used in emergency situation in 2005
- ☐ Updated in 2005
- ☐ Had plans but did not test, use, or update in 2005
- ☐ Other ⟶ Specify: _____
- ☐ Not applicable; did not have these plans in 2005

d. In 2005, how frequently did this company conduct formal vulnerability/risk assessments prior to implementing new applications, systems, or programs? Mark ■ all that apply.

- ☐ Always
- ☐ More than half the time
- ☐ Less than half the time
- ☐ When required by law
- ☐ Other ⟶ Specify: _____
- ☐ Never
- ☐ Did not implement any new applications, systems, or programs in 2005.

e. In 2005, did this company track downtime caused by any computer security incidents?

- ○ Yes
- ○ No

13532

III. TYPES OF COMPUTER SECURITY INCIDENTS

The questions in this section pertain to computer security incidents against this company, where the word "incident" refers to any unauthorized access, intrusion, breach, compromise or use of this company's computer system.

Computer security incidents may be committed by people either inside or outside the company and include computer virus, denial of service, vandalism, sabotage, embezzlement, fraud, theft of intellectual property, theft of personal or financial information, or other incidents such as hacking, spoofing, or spyware.

Please do NOT duplicate information. If an incident can be classified under multiple categories, report it under the FIRST applicable category. For example, if part of the company's computer system was deliberately damaged by means of a virus, report this under computer virus, not vandalism or sabotage.

ESTIMATES are acceptable.

5. COMPUTER VIRUS

A computer virus is a hidden fragment of computer code which propagates by inserting itself into or modifying other programs.

INCLUDE viruses, worms, Trojan horses, etc.

EXCLUDE spyware, adware, other malware, etc. Report these in 12 (Other Computer Security Incidents) on page 11.

a. In 2005, did this company intercept any computer viruses before they could infect any part of its computer systems?

○ Yes
○ No
○ Don't know

b. Did this company detect any viruses which infected any part of its computer systems in 2005?

○ Yes → **How many incidents were detected?**
   If a virus simultaneously infects a server and one or more PCs, count this as ONE INCIDENT.   Number

○ No → **(If "No", skip to 6.)**

c. Which of the following types of security technology or practices were inadequate in preventing these incidents? *Mark ■ all that apply.*

☐ Internal computer security controls	☐ Encryption
☐ External computer security controls	☐ Software vulnerability/buffer overload
☐ Anti-Virus software	☐ E-mail filters or review of e-mail logs
☐ Anti-spyware/adware software	☐ Review of system/security admin logs
☐ Biometrics	☐ Computer network/watch center
☐ One-time password generators	☐ Configuration management
☐ Passwords that must be changed	☐ Physical/environmental security
☐ Digital certificates	☐ Personnel policies
☐ Firewall	☐ Authorized access misused
☐ DMZ Host	☐ Other → Specify:
☐ Intrusion Detection System	
☐ Intrusion Protection System	☐ Don't know

d. Through which of the following were the viruses introduced into this company's networks in these incidents? *Mark ■ all that apply.*

☐ E-mail attachments
☐ Software installation
☐ Files brought in on portable media such as floppy disks, CDs, or flash drives
☐ Files downloaded from the Internet
☐ Other → Specify:
☐ Don't know

e. To which of the following organizations were these incidents reported? *Mark ■ all that apply.*

☐ Local law enforcement
☐ State law enforcement
☐ FBI (Federal Bureau of Investigation)
☐ US-CERT (United States Computer Emergency Readiness Team)
☐ Other Federal agency → Specify:
☐ CERT® Coordination Center
☐ ISAC (Information Sharing and Analysis Center)
☐ InfraGard
☐ None of the above

f. How many of these incidents were reported to the organizations specified in 5e?   Number

g. If any incidents were not reported to the organizations specified in 5e, what were the reasons? *Mark ■ all that apply.*

☐ Handled internally
☐ Reported to third party contractor providing computer security services
☐ Reported to another organization → Specify:
☐ Negative publicity
☐ Lower customer/client/investor confidence
☐ Competitor advantage
☐ Did not want data/hardware seized as evidence
☐ Did not know who to contact
☐ Incident outside jurisdiction of law enforcement
☐ Did not think to report
☐ Nothing to be gained/nothing worth pursuing
☐ Other → Specify:

13532

Page 4

III. TYPES OF COMPUTER SECURITY INCIDENTS – Continued

h. What was the relationship between the suspected offender (the person who sent or created the virus) and this company at the time of the incidents indicated in 5b? *Mark ■ all that apply.*

- ☐ Insider - someone currently (or formerly) working for this company
- ☐ Current employee
- ☐ Current contractor, vendor, temporary worker, etc.
- ☐ Former employee, contractor, vendor, temporary worker, etc.
- ☐ Outsider - someone who never worked for this company
- ☐ Domestic competitor
- ☐ Foreign competitor ⟶ Specify country: _____
- ☐ Domestic hacker
- ☐ Foreign hacker ⟶ Specify country: _____
- ☐ Other hacker (origin unknown)
- ☐ Other ⟶ Specify: _____
- ☐ Don't know

i. What was the total downtime (in hours) for each of the following due to these virus infections? ESTIMATES are acceptable.

INCLUDE downtime needed for repair.

1. Downtime of servers, routers or switches   Hours □□□□□

2. Downtime of individual PCs/workstations   Hours □□□□□
EXCLUDE network downtime reported above in item i,1.

j. How much was spent in 2005 to recover from these computer viruses? ESTIMATES are acceptable.

INCLUDE the cost - both internal and external - of diagnosis, repair, and replacement such as labor, hardware, software, etc.

	Mil.	Thou.	Dol.
$			0 0 0

EXCLUDE costs associated solely with the prevention of future incidents.

k. What other monetary losses and costs were incurred in 2005 due to these incidents? ESTIMATES are acceptable.

INCLUDE actual losses such as the value of lost information.
INCLUDE the estimated value of downtime, lost productivity,

	Mil.	Thou.	Dol.
$			0 0 0

income from lost sales, labor or fees for legal or investigative work, etc.

13532

FORM NCSS-1

III. TYPES OF COMPUTER SECURITY INCIDENTS – Continued

6. DENIAL OF SERVICE

Denial of service is the disruption, degradation, or exhaustion of an Internet connection or e-mail service that results in an interruption of the normal flow of information. Denial of service is usually caused by ping attacks, port scanning probes, excessive amounts of incoming data, etc.

EXCLUDE incidents already reported under 5 (Computer Virus) on page 3.

a. Did this company detect any incidents of denial of service (a noticeable interruption of its Internet connection or e-mail service) in 2005?

○ Yes → **How many incidents were detected?**

○ No → **(If "No", skip to 7.)**

Number

b. Which of the following types of security technology or practices were inadequate in preventing these incidents? *Mark ■ all that apply.*

☐ Internal computer security controls
☐ External computer security controls
☐ Anti-virus software
☐ Anti-spyware/adware software
☐ Biometrics
☐ One-time password generators
☐ Passwords that must be changed
☐ Digital certificates
☐ Firewall
☐ DMZ Host
☐ Intrusion Detection System
☐ Intrusion Protection System

☐ Encryption
☐ Software vulnerability/buffer overload
☐ E-mail filters or review of e-mail logs
☐ Review of system/security admin logs
☐ Computer network/watch center
☐ Configuration management
☐ Physical/environmental security
☐ Personnel policies
☐ Authorized access misused
☐ Other → Specify: _____

☐ Don't know

c. Which of the following were used, accessed, or affected in these incidents? *Mark ■ all that apply.*

☐ Local area network (LAN)
☐ Wide area network (WAN)
☐ Process control network (PCN)
☐ Virtual private network (VPN)
☐ Wireless network (e.g.,802.11)
☐ Electronic data interchange (EDI)
☐ Internet

☐ Intranet
☐ Extranet
☐ Stand-alone PCs (not on LAN)
☐ Company-owned laptop
☐ Laptop not owned by company
☐ Other → Specify: _____

☐ Don't know

d. To which of the following organizations were these incidents reported? *Mark ■ all that apply.*

☐ Local law enforcement
☐ State law enforcement
☐ FBI (Federal Bureau of Investigation)
☐ US-CERT (United States Computer Emergency Readiness Team)
☐ Other Federal agency → Specify: _____
☐ CERT® Coordination Center
☐ ISAC (Information Sharing and Analysis Center)
☐ InfraGard
☐ None of the above

e. How many of these incidents were reported to the organizations specified in 6d?

Number

f. If any incidents were not reported to the organizations specified in 6d, what were the reasons? *Mark ■ all that apply.*

☐ Handled internally
☐ Reported to third party contractor providing computer security services
☐ Reported to another organization → Specify: _____
☐ Negative publicity
☐ Lower customer/client/investor confidence
☐ Competitor advantage
☐ Did not want data/hardware seized as evidence
☐ Did not know who to contact
☐ Incident outside jurisdiction of law enforcement
☐ Did not think to report
☐ Nothing to be gained/nothing worth pursuing
☐ Other → Specify: _____

g. What was the relationship between the suspected offender and this company at the time of the incidents indicated in 6a? *Mark ■ all that apply.*

☐ Insider - someone currently (or formerly) working for this company
☐ Current employee
☐ Current contractor, vendor, temporary worker, etc.
☐ Former employee, contractor, vendor, temporary worker, etc.
☐ Outsider - someone who never worked for this company
☐ Domestic competitor
☐ Foreign competitor → Specify country: _____
☐ Domestic hacker
☐ Foreign hacker → Specify country: _____
☐ Other hacker (origin unknown)
☐ Other → Specify: _____
☐ Don't know

h. What was the total duration (in hours) of the incidents of denial of service indicated in 6a? ESTIMATES are acceptable. INCLUDE downtime needed for repairs.

Hours

i. How much was spent in 2005 to recover from these incidents of denial of service? ESTIMATES are acceptable. INCLUDE the cost - both internal and external - of diagnosis, repair, and replacement such as labor, hardware, software, etc. EXCLUDE costs associated solely with the prevention of future incidents.

Mil.	Thou.	Dol.
		0 0 0

$

j. What other monetary losses and costs were incurred in 2005 due to these incidents? ESTIMATES are acceptable. INCLUDE the estimated value of downtime, lost productivity, income from lost sales, labor or fees for legal or investigative work, etc.

Mil.	Thou.	Dol.
		0 0 0

$

13532

FORM NCSS-1

■ Page 6   ■

III. TYPES OF COMPUTER SECURITY INCIDENTS – Continued

7. ELECTRONIC VANDALISM OR SABOTAGE

Electronic vandalism or sabotage is the deliberate or malicious damage, defacement, destruction or other alteration of electronic files, data, web pages, programs, etc.

EXCLUDE incidents already reported under 5 (Computer Virus) on page 3.

EXCLUDE incidents of alteration which resulted in fraud. Report these in 9 (Fraud) on page 8.

a. Did this company detect any incidents in which files, data, web pages or any part of its computer systems were electronically vandalized or sabotaged in 2005?

○ Yes → **How many incidents were detected?** ☐☐☐☐☐☐

○ No → **(If "No", skip to 8.)**
                                       Number

b. Which of the following types of security technology or practices were inadequate in preventing these incidents? *Mark* ■ *all that apply.*

☐ Internal computer security controls ☐ Encryption
☐ External computer security controls ☐ Software vulnerability/buffer overload
☐ Anti-virus software ☐ E-mail filters or review of e-mail logs
☐ Anti-spyware/adware software ☐ Review of system/security admin logs
☐ Biometrics ☐ Computer network/watch center
☐ One-time password generators ☐ Configuration management
☐ Passwords that must be changed ☐ Physical/environmental security
☐ Digital certificates ☐ Personnel policies
☐ Firewall ☐ Authorized access misused
☐ DMZ Host ☐ Other → Specify:
☐ Intrusion Detection System
☐ Intrusion Protection System ☐ Don't know

c. Which of the following were used, accessed, or affected in these incidents? *Mark* ■ *all that apply.*

☐ Local area network (LAN) ☐ Intranet
☐ Wide area network (WAN) ☐ Extranet
☐ Process control network (PCN) ☐ Stand-alone PCs (not on LAN)
☐ Virtual private network (VPN) ☐ Company-owned laptop
☐ Wireless network (e.g.,802.11) ☐ Laptop not owned by company
☐ Electronic data interchange (EDI) ☐ Other → Specify:
☐ Internet
☐ Don't know

d. To which of the following organizations were these incidents reported? *Mark* ■ *all that apply.*

☐ Local law enforcement
☐ State law enforcement
☐ FBI (Federal Bureau of Investigation)
☐ US-CERT (United States Computer Emergency Readiness Team)
☐ Other Federal agency → Specify: _____
☐ CERT® Coordination Center
☐ ISAC (Information Sharing and Analysis Center)
☐ InfraGard
☐ None of the above

e. How many of these incidents were reported to the organizations specified in 7d? ☐☐☐☐☐☐
                                        Number

f. If any incidents were not reported to the organizations listed in 7d, what were the reasons? *Mark* ■ *all that apply.*

☐ Handled internally
☐ Reported to third party contractor providing computer security services
☐ Reported to another organization → Specify: _____
☐ Negative publicity
☐ Lower customer/client/investor confidence
☐ Competitor advantage
☐ Did not want data/hardware seized as evidence
☐ Did not know who to contact
☐ Incident outside jurisdiction of law enforcement
☐ Did not think to report
☐ Nothing to be gained/nothing worth pursuing
☐ Other → Specify: _____

g. What was the relationship between the suspected offender and this company at the time of the incidents indicated in 7a? *Mark* ■ *all that apply.*

☐ Insider - someone currently (or formerly) working for this company
☐ Current employee
☐ Current contractor, vendor, temporary worker, etc.
☐ Former employee, contractor, vendor, temporary worker, etc.
☐ Outsider - someone who never worked for this company
☐ Domestic competitor
☐ Foreign competitor → Specify country: _____
☐ Domestic hacker
☐ Foreign hacker → Specify country: _____
☐ Other hacker (origin unknown)
☐ Other → Specify: _____
☐ Don't know

h. What was the total downtime (in hours) of each of the following due to these acts of vandalism or sabotage? ESTIMATES are acceptable.

INCLUDE downtime needed for repair.
1. Downtime of company websites/ web servers   Hours ☐☐☐☐☐
2. Downtime of servers, routers or switches   Hours ☐☐☐☐☐
EXCLUDE downtime reported above in item h,1.
3. Downtime of individual PCs/workstations   Hours ☐☐☐☐☐
EXCLUDE downtime reported above in item h,1 or 2.

i. How much was spent in 2005 to recover from these incidents of vandalism or sabotage? ESTIMATES are acceptable. INCLUDE the cost - both internal and external - of diagnosis, repair, and replacement such as labor, hardware, software, etc. EXCLUDE costs associated solely with the prevention of future incidents.

	Mil.	Thou.	Dol.
$	☐☐☐	☐☐☐	0 0 0

j. What other monetary losses and costs were incurred in 2005 due to these incidents? ESTIMATES are acceptable. INCLUDE actual losses such as the value of lost information. INCLUDE the estimated value of downtime, lost productivity, income from lost sales, labor or fees for legal or investigative work, etc.

	Mil.	Thou.	Dol.
$	☐☐☐	☐☐☐	0 0 0

13532

■

III. TYPES OF COMPUTER SECURITY INCIDENTS – Continued

8. EMBEZZLEMENT

Embezzlement is the unlawful misappropriation of money or other things of value, BY THE PERSON TO WHOM IT WAS ENTRUSTED (typically an employee), for his/her own use or purpose.

INCLUDE instances in which a computer was used to wrongfully transfer, counterfeit, forge or gain access to money, property, financial documents, insurance policies, deeds, use of rental cars, various services, etc., by the person to whom it was entrusted.

a. Did this company detect any incidents in which a computer was used to commit embezzlement against this company in 2005?

○ Yes → **How many incidents were detected?**

○ No → **(If "No", skip to 9.)**

Number

b. Which of the following types of security technology or practices were inadequate in preventing these incidents? *Mark ■ all that apply.*

- ☐ Internal computer security controls
- ☐ External computer security controls
- ☐ Anti-virus software
- ☐ Anti-spyware/adware software
- ☐ Biometrics
- ☐ One-time password generators
- ☐ Passwords that must be changed
- ☐ Digital certificates
- ☐ Firewall
- ☐ DMZ Host
- ☐ Intrusion Detection System
- ☐ Intrusion Protection System
- ☐ Encryption
- ☐ Software vulnerability/buffer overload
- ☐ E-mail filters or review of e-mail logs
- ☐ Review of system/security admin logs
- ☐ Computer network/watch center
- ☐ Configuration management
- ☐ Physical/environmental security
- ☐ Personnel policies
- ☐ Authorized access misused
- ☐ Other → Specify: _____
- ☐ Don't know

c. Which of the following were used, accessed, or affected in these incidents? *Mark ■ all that apply.*

- ☐ Local area network (LAN)
- ☐ Wide area network (WAN)
- ☐ Process control network (PCN)
- ☐ Virtual private network (VPN)
- ☐ Wireless network (e.g.,802.11)
- ☐ Electronic data interchange (EDI)
- ☐ Internet
- ☐ Intranet
- ☐ Extranet
- ☐ Stand-alone PCs (not on LAN)
- ☐ Company-owned laptop
- ☐ Laptop not owned by company
- ☐ Other → Specify: _____
- ☐ Don't know

d. To which of the following official organizations were these incidents reported? *Mark ■ all that apply.*

- ☐ Local law enforcement
- ☐ State law enforcement
- ☐ FBI (Federal Bureau of Investigation)
- ☐ US-CERT (United States Computer Emergency Readiness Team)
- ☐ Other Federal agency → Specify: _____
- ☐ CERT® Coordination Center
- ☐ ISAC (Information Sharing and Analysis Center)
- ☐ InfraGard
- ☐ None of the above

e. How many of these incidents were reported to the organizations specified in 8d?

Number

f. If any incidents were not reported to the organizations specified in 8d, what were the reasons? *Mark ■ all that apply.*

- ☐ Handled internally
- ☐ Reported to third party contractor providing computer security services
- ☐ Reported to another organization → Specify: _____
- ☐ Negative publicity
- ☐ Lower customer/client/investor confidence
- ☐ Competitor advantage
- ☐ Did not want data/hardware seized as evidence
- ☐ Did not know who to contact
- ☐ Incident outside jurisdiction of law enforcement
- ☐ Did not think to report
- ☐ Nothing to be gained/nothing worth pursuing
- ☐ Other → Specify: _____

g. What was the relationship between the suspected offender and this company at the time of the incidents indicated in 8a? *Mark ■ all that apply.*

- ☐ Insider - someone currently (or formerly) working for this company
- ☐ Current employee
- ☐ Current contractor, vendor, temporary worker, etc.
- ☐ Former employee, contractor, vendor, temporary worker, etc.
- ☐ Outsider - someone who never worked for this company
- ☐ Domestic competitor
- ☐ Foreign competitor → Specify country: _____
- ☐ Domestic hacker
- ☐ Foreign hacker → Specify country: _____
- ☐ Other hacker (origin unknown)
- ☐ Other → Specify: _____
- ☐ Don't know

h. What was the dollar value of money or other things taken by embezzlement in 2005? ESTIMATES are acceptable.

$ | Mil. | Thou. | Dol. |
0 0 0

i. What other monetary losses and costs were incurred in 2005 due to these incidents? ESTIMATES are acceptable. INCLUDE the cost of diagnosis, repair and replacement such as labor, hardware, software, etc. If possible, include the estimated value of downtime, lost productivity, income from lost sales, labor or fees for legal or investigative work, etc. EXCLUDE costs associated solely with the prevention of future incidents.

$ | Mil. | Thou. | Dol. |
0 0 0

13532

FORM NCSS-1

Page 8

III. TYPES OF COMPUTER SECURITY INCIDENTS – Continued

9. FRAUD

Fraud is the intentional misrepresentation of information or identity to deceive others, the unlawful use of credit/debit card or ATM or the use of electronic means to transmit deceptive information, in order to obtain money or other things of value. Fraud may be committed by someone inside or outside the company.

INCLUDE instances in which a computer was used by someone inside or outside this company in order to defraud this company of money, property, financial documents, insurance policies, deeds, use of rental cars, various services, etc., by means of forgery, misrepresented identity, credit card or wire fraud, etc.

EXCLUDE incidents of embezzlement. Report these in 8 (Embezzlement) on page 7.

a. Did this company detect any incidents in which someone inside or outside this company used a computer to commit fraud against this company in 2005?

○ Yes → **How many incidents were detected?**

○ No → **(If "No", skip to 10.)**

Number

b. Which of the following types of security technology or practices were inadequate in preventing these incidents? *Mark ■ all that apply.*

☐ Internal computer security controls   ☐ Encryption
☐ External computer security controls   ☐ Software vulnerability/buffer overload
☐ Anti-virus software   ☐ E-mail filters or review of e-mail logs
☐ Anti-spyware/adware software   ☐ Review of system/security admin logs
☐ Biometrics   ☐ Computer network/watch center
☐ One-time password generators   ☐ Configuration management
☐ Passwords that must be changed   ☐ Physical/environmental security
☐ Digital certificates   ☐ Personnel policies
☐ Firewall   ☐ Authorized access misused
☐ DMZ Host   ☐ Other → Specify:
☐ Intrusion Detection System
☐ Intrusion Protection System   ☐ Don't know

c. Which of the following were used, accessed, or affected in these incidents? *Mark ■ all that apply.*

☐ Local area network (LAN)   ☐ Intranet
☐ Wide area network (WAN)   ☐ Extranet
☐ Process control network (PCN)   ☐ Stand-alone PCs (not on LAN)
☐ Virtual private network (VPN)   ☐ Company-owned laptop
☐ Wireless network (e.g.,802.11)   ☐ Laptop not owned by company
☐ Electronic data interchange (EDI)   ☐ Other → Specify:
☐ Internet
                                           ☐ Don't know

d. To which of the following organizations were these incidents reported? *Mark ■ all that apply.*

☐ Local law enforcement   ☐ Other Federal agency → Specify:
☐ State law enforcement
☐ FBI (Federal Bureau of Investigation)   ☐ CERT® Coordination Center
☐ US-CERT (United States Computer   ☐ ISAC (Information Sharing
   Emergency Readiness Team)   and Analysis Center)
                                           ☐ InfraGard
                                           ☐ None of the above

e. How many of these incidents were reported to the organizations specified in 9d?

Number

f. If any incidents were not reported to the organizations specified in 9d, what were the reasons? *Mark ■ all that apply.*

☐ Handled internally
☐ Reported to third party contractor providing computer security services
☐ Reported to another organization → Specify:
☐ Negative publicity
☐ Lower customer/client/investor confidence
☐ Competitor advantage
☐ Did not want data/hardware seized as evidence
☐ Did not know who to contact
☐ Incident outside jurisdiction of law enforcement
☐ Did not think to report
☐ Nothing to be gained/nothing worth pursuing
☐ Other → Specify:

g. What was the relationship between the suspected offender and this company at the time of the incidents indicated in 9a? *Mark ■ all that apply.*

☐ Insider - someone currently (or formerly) working for this company
☐ Current employee
☐ Current contractor, vendor, temporary worker, etc.
☐ Former employee, contractor, vendor, temporary worker, etc.
☐ Outsider - someone who never worked for this company
☐ Domestic competitor
☐ Foreign competitor → Specify country:
☐ Domestic hacker
☐ Foreign hacker → Specify country:
☐ Other hacker (origin unknown)
☐ Other → Specify:
☐ Don't know

h. What was the dollar value of money or other things taken by fraud in 2005? ESTIMATES are acceptable.

	Mil.	Thou.	Dol.
$			0 0 0

i. What other monetary losses and costs were incurred in 2005 due to these incidents? ESTIMATES are acceptable. INCLUDE the cost of diagnosis, repair and replacement such as labor, hardware, software, etc. If possible, include the estimated value of downtime, lost productivity, income from lost sales, labor or fees for legal or investigative work, etc. EXCLUDE costs associated solely with the prevention of future incidents.

	Mil.	Thou.	Dol.
$			0 0 0

13532

III. TYPES OF COMPUTER SECURITY INCIDENTS – Continued

10. THEFT OF INTELLECTUAL PROPERTY

Theft of intellectual property is the illegal obtaining of copyrighted or patented material, trade secrets, or trademarks including designs, plans, blueprints, codes, computer programs, software, formulas, recipes, graphics, etc., usually by electronic copying.

EXCLUDE incidents of theft of personal or financial data such as credit card or social security numbers, names and dates of birth, financial account information, etc. Report these in 11 (Theft of Personal or Financial Data) on page 10.

EXCLUDE incidents of theft of any other type of information. Report these in 12 (Other Computer Security Incidents) on page 11.

a. Did this company detect any incidents in which someone inside or outside this company used a computer to obtain intellectual property from this company in 2005?

○ Yes —▷ **How many incidents were detected?**

○ No —▷ **(If "No", skip to 11.)**

Number

b. What type of intellectual property was obtained? Mark ■ all that apply.

☐ Copyrighted material   ☐ Trade secrets

☐ Patented material   ☐ Trademarks

c. Which of the following types of security technology or practices were inadequate in preventing these incidents? Mark ■ all that apply.

☐ Internal computer security controls   ☐ Encryption

☐ External computer security controls   ☐ Software vulnerability/buffer overload

☐ Anti-virus software   ☐ E-mail filters or review of e-mail logs

☐ Anti-spyware/adware software   ☐ Review of system/security admin logs

☐ Biometrics   ☐ Computer network/watch center

☐ One-time password generators   ☐ Configuration management

☐ Passwords that must be changed   ☐ Physical/environmental security

☐ Digital certificates   ☐ Personnel policies

☐ Firewall   ☐ Authorized access misused

☐ DMZ Host   ☐ Other —▷ Specify:

☐ Intrusion Detection System

☐ Intrusion Protection System   ☐ Don't know

d. Which of the following were used, accessed, or affected in these incidents? Mark ■ all that apply.

☐ Local area network (LAN)   ☐ Intranet

☐ Wide area network (WAN)   ☐ Extranet

☐ Process control network (PCN)   ☐ Stand-alone PCs (not on LAN)

☐ Virtual private network (VPN)   ☐ Company-owned laptop

☐ Wireless network (e.g.,802.11) ☐ Laptop not owned by company

☐ Electronic data interchange (EDI) ☐ Other —▷ Specify:

☐ Internet

☐ Don't know

e. To which of the following organizations were these incidents reported? Mark ■ all that apply.

☐ Local law enforcement   ☐ Other Federal agency —▷ Specify:

☐ State law enforcement

☐ FBI (Federal Bureau of Investigation) ☐ CERT® Coordination Center

☐ US-CERT (United States Computer Emergency Readiness Team) ☐ ISAC (Information Sharing and Analysis Center)

☐ InfraGard

☐ None of the above

f. How many of these incidents were reported to the organizations specified in 10e?

Number

g. If any incidents were not reported to the organizations specified in 10e, what were the reasons? Mark ■ all that apply.

☐ Handled internally

☐ Reported to third party contractor providing computer security services

☐ Reported to another organization —▷ Specify:

☐ Negative publicity

☐ Lower customer/client/investor confidence

☐ Competitor advantage

☐ Did not want data/hardware seized as evidence

☐ Did not know who to contact

☐ Incident outside jurisdiction of law enforcement

☐ Did not think to report

☐ Nothing to be gained/nothing worth pursuing

☐ Other —▷ Specify:

h. What was the relationship between the suspected offender and this company at the time of the incidents indicated in 10a? Mark ■ all that apply.

☐ Insider - someone currently (or formerly) working for this company

☐ Current employee

☐ Current contractor, vendor, temporary worker, etc.

☐ Former employee, contractor, vendor, temporary worker, etc.

☐ Outsider - someone who never worked for this company

☐ Domestic competitor

☐ Foreign competitor —▷ Specify country:

☐ Domestic hacker

☐ Foreign hacker —▷ Specify country:

☐ Other hacker (origin unknown)

☐ Other —▷ Specify:

☐ Don't know

i. What was the dollar value of intellectual property taken by theft in 2005? ESTIMATES are acceptable.

Mil.	Thou.	Dol.
		0 0 0

$

j. What other monetary losses and costs were incurred in 2005 due to these incidents? ESTIMATES are acceptable. INCLUDE the cost of diagnosis, repair and replacement such as labor, hardware, software, etc. If possible, include the estimated value of downtime, lost productivity, income from lost sales, labor or fees for legal or investigative work, etc. EXCLUDE costs associated solely with the prevention of future incidents.

Mil.	Thou.	Dol.
		0 0 0

$

k. How many of the incidents indicated in 10a involved unlicensed use or copying (piracy) of digital products which this company developed for resale?

Number

13532

FORM NCSS-1

■ Page 10

III. TYPES OF COMPUTER SECURITY INCIDENTS – Continued

11. THEFT OF PERSONAL OR FINANCIAL INFORMATION

Theft of personal or financial information is the illegal obtaining of information that could potentially allow someone to use or create accounts under another name (individual, business, or some other entity). Personal information includes names, dates of birth, social security numbers, etc. Financial information includes credit/debit/ATM card, account, or PIN numbers, etc.

EXCLUDE incidents of theft of intellectual property such as copyrights, patents, trade secrets, and trademarks. Report these in 10 (Theft of Intellectual Property) on page 9.

EXCLUDE incidents of theft of any other type of information. Report these in 12 (Other Computer Security Incidents) on page 11.

a. Did this company detect any incidents in which someone inside or outside this company used a computer to obtain personal or financial information from this company in 2005?

○ Yes → **How many incidents were detected?**

○ No → **(If "No", skip to 12.)**

Number

b. What type of personal or financial information was obtained? *Mark* ■ *all that apply.*

☐ Names or dates of birth   ☐ Debit or ATM card numbers

☐ Social security numbers   ☐ Account or PIN numbers

☐ Credit card numbers   ☐ Other → Specify:

c. Which of the following types of security technology or practices were inadequate in preventing these incidents? *Mark* ■ *all that apply.*

☐ Internal computer security controls   ☐ Encryption

☐ External computer security controls   ☐ Software vulnerability/buffer overload

☐ Anti-virus software   ☐ E-mail filters or review of e-mail logs

☐ Anti-spyware/adware software   ☐ Review of system/security admin logs

☐ Biometrics   ☐ Computer network/watch center

☐ One-time password generators   ☐ Configuration management

☐ Passwords that must be changed   ☐ Physical/environmental security

☐ Digital certificates   ☐ Personnel policies

☐ Firewall   ☐ Authorized access misused

☐ DMZ Host   ☐ Other → Specify:

☐ Intrusion Detection System

☐ Intrusion Protection System   ☐ Don't know

d. Which of the following were used, accessed, or affected in these incidents? *Mark* ■ *all that apply.*

☐ Local area network (LAN)   ☐ Intranet

☐ Wide area network (WAN)   ☐ Extranet

☐ Process control network (PCN)   ☐ Stand-alone PCs (not on LAN)

☐ Virtual private network (VPN)   ☐ Company-owned laptop

☐ Wireless network (e.g.,802.11)   ☐ Laptop not owned by company

☐ Electronic data interchange (EDI)   ☐ Other → Specify:

☐ Internet

☐ Don't know

e. To which of the following organizations were these incidents reported? *Mark* ■ *all that apply.*

☐ Local law enforcement   ☐ CERT® Coordination Center

☐ State law enforcement   ☐ ISAC (Information Sharing and Analysis Center)

☐ FBI (Federal Bureau of Investigation Emergency Readiness Team)   ☐ InfraGard

☐ US-CERT (United States Computer   ☐ None of the above

☐ Other Federal agency → Specify:

f. How many of these incidents were reported to the organizations specified in 11e?

Number

g. If any incidents were not reported to the organizations specified in 11e, what were the reasons? *Mark* ■ *all that apply.*

☐ Handled internally

☐ Reported to third party contractor providing computer security services

☐ Reported to another organization → Specify:

☐ Negative publicity

☐ Lower customer/client/investor confidence

☐ Competitor advantage

☐ Did not want data/hardware seized as evidence

☐ Did not know who to contact

☐ Incident outside jurisdiction of law enforcement

☐ Did not think to report

☐ Nothing to be gained/nothing worth pursuing

☐ Other → Specify:

h. What was the relationship between the suspected offender and this company at the time of the incidents indicated in 11a? *Mark* ■ *all that apply.*

☐ Insider - someone currently (or formerly) working for this company

☐ Current employee

☐ Current contractor, vendor, temporary worker, etc.

☐ Former employee, contractor, vendor, temporary worker, etc.

☐ Outsider - someone who never worked for this company

☐ Domestic competitor

☐ Foreign competitor → Specify country:

☐ Domestic hacker

☐ Foreign hacker → Specify country:

☐ Other hacker (origin unknown)

☐ Other → Specify:

☐ Don't know

i. What was the dollar value of personal or financial information taken by theft in 2005? ESTIMATES are acceptable.

Mil.	Thou.	Dol.
		0 0 0

$

j. What other monetary losses and costs were incurred in 2005 due to these incidents? ESTIMATES are acceptable. INCLUDE the cost of diagnosis, repair and replacement such as labor, hardware, software, etc. If possible, include the estimated value of downtime, lost productivity, income from lost sales, labor or fees for legal or investigative work, etc. EXCLUDE costs associated solely with the prevention of future incidents.

Mil.	Thou.	Dol.
		0 0 0

$

13532

III. TYPES OF COMPUTER SECURITY INCIDENTS – Continued

12. OTHER COMPUTER SECURITY INCIDENTS

INCLUDE all other computer security incidents involving this company's computer networks—such as hacking, sniffing, spyware, theft of other information—regardless of whether damage or losses were sustained as a result.

EXCLUDE incidents already reported in this survey.

a. Did this company detect any other computer security incidents in 2005?

○ Yes → **How many incidents were detected?**

○ No → **(If "No", skip to 13.)**

Number

b. What other types of computer security incidents were detected in 2005? *Mark ■ all that apply.*

☐ Hacking   ☐ Spyware, keystroke logging

☐ Spoofing   ☐ Adware

☐ Phishing   ☐ Other malware

☐ Sniffing   ☐ Theft of information not already reported in 10 or 11 on pages 8 or 9 → Please describe:

☐ Pinging   _____

☐ Scanning   ☐ Other → Please describe: _____

c. Which of the following types of security technology or practices were inadequate in preventing these incidents? *Mark ■ all that apply.*

☐ Internal computer security controls   ☐ Encryption

☐ External computer security controls   ☐ Software vulnerability/buffer overload

☐ Anti-virus software   ☐ E-mail filters or review of e-mail logs

☐ Anti-spyware/adware software   ☐ Review of system/security admin logs

☐ Biometrics   ☐ Computer network/watch center

☐ One-time password generators   ☐ Configuration management

☐ Passwords that must be changed   ☐ Physical/environmental security

☐ Digital certificates   ☐ Personnel policies

☐ Firewall   ☐ Authorized access misused

☐ DMZ Host   ☐ Other → Specify: _____

☐ Intrusion Detection System   _____

☐ Intrusion Protection System   ☐ Don't know

d. Which of the following were used, accessed, or affected in these incidents? *Mark ■ all that apply.*

☐ Local area network (LAN)   ☐ Intranet

☐ Wide area network (WAN)   ☐ Extranet

☐ Process control network (PCN)   ☐ Stand-alone PCs (not on LAN)

☐ Virtual private network (VPN)   ☐ Company-owned laptop

☐ Wireless network (e.g.,802.11)   ☐ Laptop not owned by company

☐ Electronic data interchange (EDI) ☐ Other → Specify:

☐ Internet   _____

                                   ☐ Don't know

e. To which of the following organizations were these incidents reported? *Mark ■ all that apply.*

☐ Local law enforcement   ☐ Other Federal agency → Specify:

☐ State law enforcement   _____

☐ FBI (Federal Bureau of Investigation) ☐ CERT® Coordination Center

☐ US-CERT (United States Computer Emergency Readiness Team) ☐ ISAC (Information Sharing and Analysis Center)

                                   ☐ InfraGard

                                   ☐ None of the above

f. How many of these incidents were reported to the organizations specified in 12e?

Number

g. If any incidents were not reported to the organizations listed in 12e, what were the reasons? *Mark ■ all that apply.*

☐ Handled internally

☐ Reported to third party contractor providing computer security services

☐ Reported to another organization → Specify: _____

☐ Negative publicity

☐ Lower customer/client/investor confidence

☐ Competitor advantage

☐ Did not want data/hardware seized as evidence

☐ Did not know who to contact

☐ Incident outside jurisdiction of law enforcement

☐ Did not think to report

☐ Nothing to be gained/nothing worth pursuing

☐ Other → Specify: _____

h. What was the relationship between the suspected offender and this company at the time of the incidents indicated in 12a? *Mark ■ all that apply.*

☐ Insider - someone currently (or formerly) working for this company

☐ Current employee

☐ Current contractor, vendor, temporary worker, etc.

☐ Former employee, contractor, vendor, temporary worker, etc.

☐ Outsider - someone who never worked for this company

☐ Domestic competitor

☐ Foreign competitor → Specify country: _____

☐ Domestic hacker

☐ Foreign hacker → Specify country: _____

☐ Other hacker (origin unknown)

☐ Other → Specify: _____

☐ Don't know

i. If any, what was the total downtime (in hours) of each of the following due to these other computer security incidents? ESTIMATES are acceptable.

INCLUDE downtime needed for repair.

1. Downtime of company websites/ web servers   Hours

2. Downtime of servers, routers or switches   Hours
EXCLUDE downtime reported above in item i,1.

3. Downtime of individual PCs/workstations   Hours
EXCLUDE downtime reported above in item i,1 or 2.

j. How much was spent in 2005 to recover from these other computer security incidents? ESTIMATES are acceptable.
INCLUDE the cost - both internal and external - of diagnosis, repair, and replacement such as labor, hardware, software, etc.
EXCLUDE costs associated solely with the prevention of future incidents.

Mil.	Thou.	Dol.
$		0 0 0

k. What other monetary losses and costs were incurred in 2005 due to these incidents?
ESTIMATES are acceptable.
INCLUDE actual losses such as the value of lost information.
INCLUDE the estimated value of downtime, lost productivity, income from lost sales, abor or fees for legal or investigative work, etc.

Mil.	Thou.	Dol.
$		0 0 0

13532

Page 12

IV. OTHER TRENDS IN COMPUTER SECURITY

13. In 2005, did this company detect any computer security breaches that resulted from information obtained from a stolen laptop computer?

○ Yes ⟶ **How many incidents were detected?**

○ No

Number

14. In 2005, was the overall number of computer security incidents detected by this company more, less or about the same compared to the number detected in 2004 regardless of whether damage or losses were sustained as a result? *Mark ● only one.*

○ More in 2005

○ Less in 2005

○ About the same

○ Don't know

15. In 2005, did this company have a separate insurance policy or rider to cover losses due specifically to computer security breaches?

○ Yes

○ No

○ Don't know

16. In 2005, what percentage of this company's business was transacted over the Internet, Intranet, Extranet, EDI, etc.?
ESTIMATES are acceptable.
INCLUDE any transaction completed over a computer-mediated network that involves the transfer of ownership or rights to use goods or services. For example, taking orders for merchandise or services, transferring information or rights, paying accounts, etc.

%

V. COMPANY INFORMATION

17. In 2005, which of the following Internet services, if any, did this company provide to other companies or individuals as its PRIMARY line of business? *Mark ■ all that apply.*

☐ Internet Service Provider (ISP)

☐ Web Search Portal

☐ Other Internet service ⟶ Specify: _____

☐ None of the above

18 a. What were the total operating revenue, sales, and/or receipts for this company in 2005?
ESTIMATES are
$

Bil.	Mil.	Thou.	Dol.
			0 0 0

b. What percentage of this total was derived from e-commerce?
ESTIMATES are acceptable.
INCLUDE any transaction completed over a computer-mediated network that involves the transfer of ownership or rights to use goods or services. For example, taking orders for merchandise or services, transferring information or rights, paying accounts, etc.

%

19. What was the total number of employees on this company's payroll for the pay period which includes March 12, 2005?
ESTIMATES are acceptable.
Count EACH part-time employee as one.
EXCLUDE contractors, vendors, leased and temporary employees.

Number

20. Does the information reported in this survey cover calendar year 2005, fiscal year 2005 or some other time period?

○ Calendar year 2005

○ Fiscal year 2005 or some other time period ⟶ Specify period covered:

FROM: | Month | Year | TO: | Month | Year |

21. Does the information reported in this survey include this company or does it include this company and some or all of its subsidiaries?
For this survey, subsidiary means a company in which this company has more than 50% ownership, or in which this company has the power to direct or cause the direction of management and policies.

○ Information includes this company only - company has no subsidiaries, or responses exclude subsidiaries

○ Information includes this company and some or all of its subsidiaries - **How many subsidiaries were included?**

Number

13532

Page 13

(Tear off sheet - identifying information will be separated from survey responses upon receipt by RAND.)

CONTACT INFORMATION

Person to contact regarding this report:

Name

Title

Company Name

Phone   Ext.
(   )   -

Fax
(   )   -

E-mail address

Please list subsidiaries included in this report:

REMARKS

(Please use this space or a separate sheet of paper for any explanations that may be essential in understanding your reported data.)

13532

FORM NCSS-1

U.S. DEPARTMENT OF JUSTICE
BUREAU OF JUSTICE STATISTICS
In partnership with the
U.S. DEPARTMENT OF HOMELAND SECURITY
NATIONAL CYBER SECURITY DIVISION

2005 NATIONAL COMPUTER SECURITY SURVEY INSTRUCTIONS

PURPOSE OF THE SURVEY

The purpose of this survey is to collect information about the nature and extent of computer security incidents experienced by businesses located in the U.S. The data you report will provide information on the impact of computer crime on businesses.

Specifically, data from the 2005 National Computer Security Survey will provide information on the frequency and types of crime involving computers, the monetary losses sustained as a result of computer crime, and the cost of computer security.

LEGAL AUTHORITY AND CONFIDENTIALITY

Your report is confidential by law (P.L. 107-347, Title V and 44 U.S.C. § 3501 note). It may be seen only by persons certified to uphold the confidentiality of this information and used only for statistical purposes from which no firm may be identified. The law also prohibits the sharing of your data with other agencies, exempts the information you provide from requests made under the Freedom of Information Act, and ensures that your responses are immune from legal process.

BURDEN HOUR ESTIMATE

Respondents are not required to respond to any information collection unless it displays a valid approval number from the Office of Management and Budget. Public reporting burden for this collection of information is estimated to vary from 45 minutes to 3 hours per response, with an average of 1½ hours per response, including the time for reviewing instructions, searching existing data sources, gathering and maintaining the data needed and completing and reviewing the collection of information. Send comments regarding this burden estimate or any other aspect of this collection of information, including suggestions for reducing this burden, to Bureau of Justice Statistics, National Computer Security Survey, Washington, DC 20531; and to the Office of Management and Budget, OMB No. 1121-0301, Washington, DC 20503.

GENERAL INSTRUCTIONS

Survey Scope – This survey collects computer security data for companies, organizations and associations operating within the United States. **Information for business-related activities of religious organizations, nonprofit organizations and organizations that are government owned but privately operated should be included.**

Reporting Entity – Report computer security data for all **domestic operations** of your company, including all divisions and locations. A company is a business, service or membership organization consisting of one or more establishments under common ownership or control. **Do not report for subsidiary companies that your company may hold, as they may be surveyed separately.** For this survey, subsidiary means a

company in which this company has more than 50% ownership, or in which this company has the power to direct or cause the direction of management and policies. *Use figures that include subsidiaries only if figures that exclude subsidiaries are not available.* For purposes of this survey, exclude data for Puerto Rico, the Virgin Islands and U.S. Territories. If you are unable to consolidate records for the entire company minus subsidiaries or have reporting questions, please call **1–800–734-5399.**

How to Report Dollar Figures – Dollar figures should be **rounded** to thousands of dollars.

For example, if the figure is $1,023,528.79, enter:

Mil.	Thou.	Dol.
$ 1	0 2 4	

If the figure is less than $500.00, enter:

Mil.	Thou.	Dol.
$	0	

Estimates are acceptable – The data requested on the National Computer Security Survey may not correspond to your company's records. If you cannot answer a question from your company records, please provide a carefully prepared estimate.

Reporting Period – Report data for calendar year 2005. If you cannot provide data on a calendar year basis, fiscal year 2005 data are acceptable. If this company was not in operation for the full year, report for the period of time it was in operation. Indicate in Question 20, Report Period, the exact dates the data represent if they are not for the calendar year.

Additional Forms – Photocopies of this form are acceptable. If you require additional forms, contact us at the toll-free number, e-mail address, or business address provided below.

Filing the Report Form – Return your completed form in the pre-addressed envelope. If you are not using the pre-addressed envelope, return it to the address provided at the bottom of this page or fax it to 1–877-814-6673.

RAND Corporation
Survey Research Group
1776 Main Street
P.O. Box 2138
Santa Monica, CA 90407-2138

Direct any **QUESTIONS** regarding this form to:

Toll-free Number: 1-800-734-5399
FAX Number: 1-877-814-6673
E-mail: ncss@rand.org

GLOSSARY OF TERMS

Adware – A software application that automatically displays advertisements, typically in the form of pop-up windows. Adware sometimes includes spyware.

Anti-spyware/adware software – A utility that looks for spyware and/or adware and alerts the user to any that are found.

Anti-virus software – A utility that looks for viruses and alerts the user to any that are found.

Biometrics – Methods of generating authentication information for a person by digitizing measurements of a physical characteristic, such as a fingerprint, a hand shape, a retinal pattern, a speech pattern (voice print), or handwriting.

Business continuity plan for computer systems – The procedure an organization uses to maintain essential functions during and after a disaster, such as a dual back-up system at a separate physical location. It seeks to ensure the uninterrupted provision of mission-critical functions. It often includes a disaster recovery plan.

Company laptops – Any laptop computer issued by this company, whether owned or leased.

Computer/network watch center – The location from which control is exercised over a communications network, usually either telephony or Internet, though sometimes also that of a public utility. It is sometimes also the location containing many or all of the primary servers and other equipment that runs an internet service provider. This center is also where the technicians that maintain the servers, develop new software, and troubleshoot outages are located.

Configuration management – The management of security features and assurances through control of changes made to hardware, software, firmware, documentation, test fixtures, and test documentation of an automated information system, throughout the development and operational life of a system. Includes Source Code Management or revision control. The control of changes—including the recording thereof—that are made to the hardware, software, firmware, and documentation throughout the system lifecycle.

Corporate policy on computer security – A defined set of practices and guidelines established by the organization to deal with issues involving computer security. Such practices and guidelines can encompass the responsibilities of both the organization and its employees. Employees have been made aware of this policy.

Digital certificates – An attachment to an electronic message used for security purposes. The most common use of a digital certificate is to verify that a user sending a message is who he or she claims to be, and to provide the receiver with the means to encode a reply.

Disaster recovery plan for computer systems – A procedure to restore an organization's mission-critical functions after, and to minimize the effects of, a major interruption of computer services. It includes procedures for reporting specific types of problems to designated personnel, repairing or replacing damaged systems, etc.

DMZ Host – A small network that acts as a "neutral zone" between a company's internal network and an external network such as the Internet. A DMZ host is usually inserted behind or between firewalls.

Electronic Data Interchange (EDI) – A proprietary electronic system used for exchanging business data over a computer network.

E-mail logs or filters – E-mail logs keep track of incoming/outgoing messages, including the sender and the recipient. Filters are an automated method of searching the content of e-mail for words, viruses, or misuse of computer resources.

Encryption – The translation of data into a format that requires a code to restore it to the original format. To read an encrypted file, you must have access to a secret key or password that allows you to decrypt it.

Equipment decommissioning – A procedure used for removing computer equipment from active use within an information system or network. This involves changing settings within the system to reflect their absence, and the removal of all sensitive information from the computer equipment, particularly from hard drives and other media.

External computer security controls – Hardware, software, and/or company policies and practices limiting the access of outsiders to the company's computer systems and networks.

Extranet – A network that uses Internet/Intranet technology to make information available to authorized outsiders. It allows businesses to securely share information with selected suppliers, partners, customers, or other businesses.

Firewall – Hardware and/or software designed to prevent unauthorized access to or from a private network, particularly networks with Internet or Intranet connectivity.

Formal computer security audit standards – An established or authoritative set of criteria used to review computer security systems.

Hacker – An unauthorized person who cracks a computer system or exceeds authorized access for malicious intent or for the thrill of the challenge.

Hard-wired telecommunication lines – Telecommunication lines that are copper or fiber-optic and stationary, as opposed to wireless.

Identification of company's critical assets – Determining the critical functions that the organization performs, and the assets (such as information and telecommunication systems) upon which those functions are vitally dependent. Those critical assets are ones for which special security and reliability measures should be focused.

Insurance covering computer security breaches – This type of insurance specifically covers losses due to computer break-in exposures, usually in a separate policy or rider. The coverage is typically called network security liability, e-commerce liability, Internet security liability, or hacker insurance.

Internal computer security controls – Hardware, software, and/or company policies and practices limiting the access of insiders—employees, contractors, vendors, etc.—to the company's computer systems or networks. These controls may vary by department and/or employee function.

Internet – Inter-connected networks linking millions of computers globally. Users can access information and applications from other computers and communicate with other users.

Intranet – An internal network similar to the Internet but surrounded by a firewall to prevent access from users outside the company, organization, or facility.

Intrusion detection system – An intrusion detection system examines all inbound and outbound network activity and identifies suspicious patterns that may signal a network or system attack from someone attempting to break into or compromise a system.

Intrusion/penetration testing of computer security – A method of evaluating the security of a computer system and identifying its vulnerabilities by attempting to circumvent or override system security through the simulation of an attack by a malicious actor.

Intrusion protection system – A suite of access control tools used to protect computers from exploitation. Intrusion protection systems may also act at the host level to deny potentially malicious activity.

Local area network (LAN) – A computer network that spans a small area such as a single building or group of buildings.

Malware – Malicious software or code developed to serve a harmful purpose. Specific types of malware include viruses, worms, Trojan horses, spyware, and adware.

Misuse of computers by employees (Internet, e-mail, etc.) – The improper use of company computer resources by employees, such as using the company's computer resources for personal gain, sending personal or improper e-mail, abusing Internet privileges, loading unlicensed software, etc.

Non-company laptop – Any laptop computer not issued by this company (e.g., belonging to a consultant, vendor, contractor, etc.).

One-time password generators (smart cards, tokens, keys) – A "one-time password generator" is an authentication device such as a one-time token which randomly changes all or part of the user's password, typically every minute, so that the same password is never used more than once. This technique counters the threat of a replay attack that uses passwords captured by spyware, wiretapping, or other means of hacking.

Passwords that must be changed periodically – A simple authentication technique in which each password is used repeatedly for a specific period of time to verify an identity.

Periodic computer security audits – Reviews conducted periodically by the company's security office. For example, the company's strike team might simulate computer security situations and then evaluate how the company performed.

Phishing – The creation and use of fraudulent but legitimate-looking e-mails and web sites to obtain users' personal and financial account information for criminal purposes.

Physical/environmental security (e.g., limited physical access, sprinklers) – Security measures focused on limiting physical access to critical organization assets, and protection of those assets from physical malicious attacks (e.g., explosions) or natural disasters (earthquakes, fire, flood).

Pinging – A basic test of whether a particular host is operating properly and is reachable on the network from the testing host by sending a special packet of information and awaiting its response. Malicious use includes flooding the Internet with ping requests attempting to locate new hosts to infect, causing problems to routers across the Internet.

Piracy – see Unlicensed use or copying.

Process control network (PCN) – A network with an automated control of a process, such as a manufacturing process or assembly line. It is used extensively in industrial operations, such as oil refining, chemical processing, and electrical generation. It uses analog devices to monitor real-world signals and digital computers to do the analysis and controlling. It makes extensive use of analog/digital, digital/analog conversion.

Publicly accessible website WITH e-commerce capabilities – E-commerce capabilities refer to the ability of this company's customers or suppliers to effect transactions via computer networks. Such transactions commit the company and the customer/supplier to an exchange, though they do not necessarily include making payment associated with the commitment. For example, if a customer orders products via a website with payment made by check at a later date, this is an e-commerce transaction.

Regular review of system administrative logs – Reviewing system administrative logs on a regular basis to detect suspicious activity beyond normal daily activity.

Remote dial-in access – Refers to using devices and other resources that are not connected directly to a workstation to connect to another computer device. Do not include network access through the Internet.

Scanning – A method of searching for open ports by sending packets or requests for information.

Server – A computer or device on a network that manages network resources. For example, a file server is a computer and storage device dedicated to storing files. A print server is a computer that manages one or more printers. A network server is a computer that manages network traffic.

Sniffing – Packet sniffing is a form of wire-tap applied to computer networks instead of phone networks. Traffic on a network segment passes by all hosts attached to that segment. Ethernet cards have a filter that prevents the host machine from seeing traffic addressed to other stations. Sniffing programs turn off the filter, and thus see everyone's traffic.

Spoofing – The creation of TCP/IP packets using someone else's IP address. A "spoofed" IP address is therefore misleading regarding the true source of an Internet message packet.

Spyware – Software that surreptitiously monitors the user and transmits the information to a third party. Some spyware can intercept or take partial control of a computer's operation. Spyware differs from viruses and worms in that it does not usually self-replicate.

Stand-alone PCs (not on LAN) – Computers that are not connected to company networks, such as a stand-alone workstation. For the purposes of this survey, a stand-alone computer may have Internet access.

System administrative logs – Logs which document details of access to computer systems, such as who logged in, which parts of the system were accessed, and when the user logged in and out.

Training employees in computer security practices – Training session(s) designed to educate employees on issues dealing with computer security and the employee's role in following the organization's computer security practices.

Trojan horse – A program that overtly does one thing while covertly doing another.

Unlicensed use or copying (piracy) of digital products developed for resale – The unauthorized copying or use of digital products — such as software, music, or motion pictures — which the company developed or for which it holds the copyright. Report unauthorized copying or use of other software by employees under "Misuse of computers by employees (Internet, e-mail, etc.)."

Virtual private network (VPN) – A network that is constructed by using public wires to connect nodes. For example, systems that allow you to create networks using the Internet as the medium for transporting data. These systems use encryption and other security mechanisms to ensure that only authorized users can access the network.

Virus – A hidden fragment of computer code which propagates by inserting itself into or modifying other programs.

Vulnerability/risk assessment – Assessment of threats to, impacts on, and vulnerabilities of information and information-processing facilities and the likelihood of their occurrence.

Wide area network (WAN) – A computer network that spans a large geographical area. Usually, a WAN consists of two or more LANs.

Wireless networks (e.g., 802.11) – A type of LAN that uses high-frequency radio waves or lasers rather than wires to communicate between nodes. 802.11 refers to a family of specifications for an over-the-air interface between a wireless client and a base station or between two wireless clients.

Wireless access to e-mail, Internet and/or this company's other networks – Wireless access refers to the use of a device or system that will enable access to a network to which it is not physically connected. For example, access via a cellular or digital phone, some personal digital assistants (PDAs), some laptop computers, thin client, broadband, etc.

Worm – A self-replicating computer program, similar to a computer virus. A virus attaches itself to, and becomes part of, another executable program; however, a worm is self-contained and does not need to be part of another program to propagate itself. They are often designed to exploit the file transmission capabilities found on many computers.

Bibliography

Collins, Linda M., Joseph L. Schafer, and Chi-Ming Kam, "A Comparison of Inclusive and Restrictive Strategies in Modern Missing Data Procedures," *Psychological Methods*, Vol. 6, No. 4, December 2001, pp. 330–351.

Dillman, Don A., *Mail and Internet Surveys: The Tailored Design Method*, 2nd ed., Hoboken, N.J.: Wiley, 2007.

Dun and Bradstreet, *The DUNSRight Process: The Power Behind Quality Information*, presentation to the RAND Corporation, May 2004.

International Organization for Standards, "Information Technology: Security Techniques—Code of Practice for Information Security Management," standard 17799:2005.

ISO—*see* International Organization for Standards.

National Science Foundation, Division of Science Resources Statistics, *Information Technology Innovation Survey: Fall 2001*, Arlington, Va., 2004. As of March 18, 2008:
http://www.nsf.gov/statistics/nsf04305/

NSF—*see* National Science Foundation.

Rantala, Ramona R., *Cybercrime Against Businesses*, Washington, D.C.: U.S. Department of Justice, Office of Justice Programs, NCJ200639, March 2004. As of March 23, 2008:
http://purl.access.gpo.gov/GPO/LPS77521

Rubin, Donald B., "Inference and Missing Data," *Biometrika*, Vol. 63, No. 3, December 1976, pp. 581–592.

Sand, Kelly L., Sarah K. Cotton, Robert H. Anderson, Daniela Golinelli, Rosalind Lewis, and Lois M. Davis, *The National Computer Security Survey Cognitive Interviewing Report*, Santa Monica, Calif.: RAND Corporation, 2005. Not cleared for open publication.

Schafer, Joseph L., and John W. Graham, "Missing Data: Our View of the State of the Art," *Psychological Methods*, Vol. 7, No. 2, June 2002, pp. 147–177.